THE CHEMISTRY BOOK

WORKBOOK

Nicholas Stansbie
Alan Knight
Brett Steeples
Sarah Windsor

The Chemistry Book Units 1 & 2
1st Edition
Nicholas Stansbie
Brett Steeples
Alan Knight
Sarah Windsor

Contributing authors: Bob Bucat, Anna Davis, Anne Disney, Suzanne Farr, Von Hayes, George Hook, Elizabeth McKenna, Deb Smith and Rachel Whan

Publishing editor: Rachel Ford
Project editor: Simon Tomlin
Editor: Kirstie Irwin
Cover design: Chris Starr (MakeWork) Text design: Watershed Design
Art direction: Petrina Griffin
Cover image: iStock.com
Permissions researcher: Debbie Gallagher
Production controller: Karen Young
Typeset by: MPS Limited
Any URLs contained in this publication were checked for currency during the production process. Note, however, that the publisher cannot vouch for the ongoing currency of URLs.

For product information and technology assistance,
in Australia call **1300 790 853**;
in New Zealand call **0800 449 725**

For permission to use material from this text or product, please email
aust.permissions@cengage.com

ISBN 978 0 17 041239 1

Cengage Learning Australia
Level 7, 80 Dorcas Street
South Melbourne, Victoria Australia 3205

Cengage Learning New Zealand
Unit 4B Rosedale Office Park
331 Rosedale Road, Albany, North Shore 0632, NZ

For learning solutions, visit **cengage.com.au**

Printed in Singapore by 1010 Printing Group Limited.
1 2 3 4 5 6 7 22 21 20 19 18

CONTENTS

9780170412391

TOPIC 3: CHEMICAL REACTIONS – REACTANTS, PRODUCTS AND ENERGY CHANGE

9780170412391

HOW TO USE THIS BOOK

Learning

The learning section is a summary of the key knowledge and skills. This summary can be used to create mind maps, to write short summaries and as a check list.

Revision

This section is a series of structured activities to help consolidate the knowledge and skills acquired in class.

Evaluation

The evaluation section is in the style of a practice exam to test and evaluate the acquisition of knowledge and skills.

Practice exam

A tear-out exam helps to facilitate preparing and practicing for external exams.

ABOUT THE AUTHORS

Nicholas Stanbsie

Nicholas is an experienced science teacher and teacher of senior chemistry in the UK and in Australia. Nicholas' dedication to quality teaching and learning has enabled him to lead the *QScience Chemistry* team in the development of this text.

Professor Alan E.W. Knight

Professor Alan E.W. Knight spent more than 35 years as a research scientist in the field of chemical physics. Alan's research interests included molecular spectroscopy, structure and bonding, reaction dynamics and molecular quantum theory, with applications in atmospheric chemistry and astrophysics. His philosophy of education is based on building a deep and lasting understanding of science.

Brett Steeples

Brett was a research chemist and lecturer at Manchester University before becoming a chemistry teacher. Brett has worked on the moderation panel and is a member of the chemistry state review panel.

Dr Sarah Windsor

Dr Sarah Windsor is an associate lecturer, science, at the University of the Sunshine Coast. Sarah's research interests include analytical chemistry and education and she has brought her extensive knowledge in both of those areas to the development of *QScience Chemistry*.

Some of the material in *The Chemistry Book Units 1 & 2* has been taken from or adapted from the following publications:
Nelson Chemistry for the Australian Curriculum Units 1 & 2 NelsonNet material written by Deb Smith, Anna Davis, Anne Disney, Von Hayes, Rachel Wan, Suzanne Farr, Elizabeth McKenna and George Hook.
Nelson Chemistry for the Australian Curriculum Units 3 & 4 NelsonNet material written by: Bob Bucat and Rachel Whan.

SYLLABUS REFERENCE GRID

UNITS AND TOPICS	*THE CHEMISTRY BOOK UNITS 1 & 2*
UNIT 1: CHEMICAL FUNDAMENTALS – STRUCTURES, PROPERTIES AND REACTIONS	
Topic 1: Properties and structures of atoms	
Periodic table and trends	Chapter 1
Atomic structure	Chapter 2
Introduction to bonding	Chapter 3
Isotopes	Chapter 4
Analytical techniques	Chapter 5
Topic 2: Properties and structure of materials	
Compounds and mixtures	Chapter 6
Bonding and properties	Chapter 7
Topic 3: Chemical reactants – reactants, products and energy change	
Chemical reactions	Chapter 8
Exothermic and endothermic reactions	Chapter 9
Measurement uncertainty and error	Chapter 10
Fuels	Chapter 11
Mole concept and law of conservation of mass	Chapter 12
UNIT 2: MOLECULAR INTERACTIONS AND REACTIONS	
Topic 1: Intermolecular forces and gases	
Intermolecular forces and gases	Chapter 13
Chromatography techniques	Chapter 14
Gases	Chapter 15
Topic 2: Aqueous solutions and acidity	
Aqueous solutions and molarity	Chapter 16
Identifying ions in solution	Chapter 17
Solubility	Chapter 18
pH	Chapter 19
Reaction of acids	Chapter 20
Topic 3: Rates of chemical reaction	
Rates of reactions	Chapter 21

UNIT ONE

CHEMICAL FUNDAMENTALS – STRUCTURES, PROPERTIES AND REACTIONS

- Topic 1: Properties and structure of atoms
- Topic 2: Properties and structure of materials
- Topic 3: Chemical reactions – reactants, products and energy change

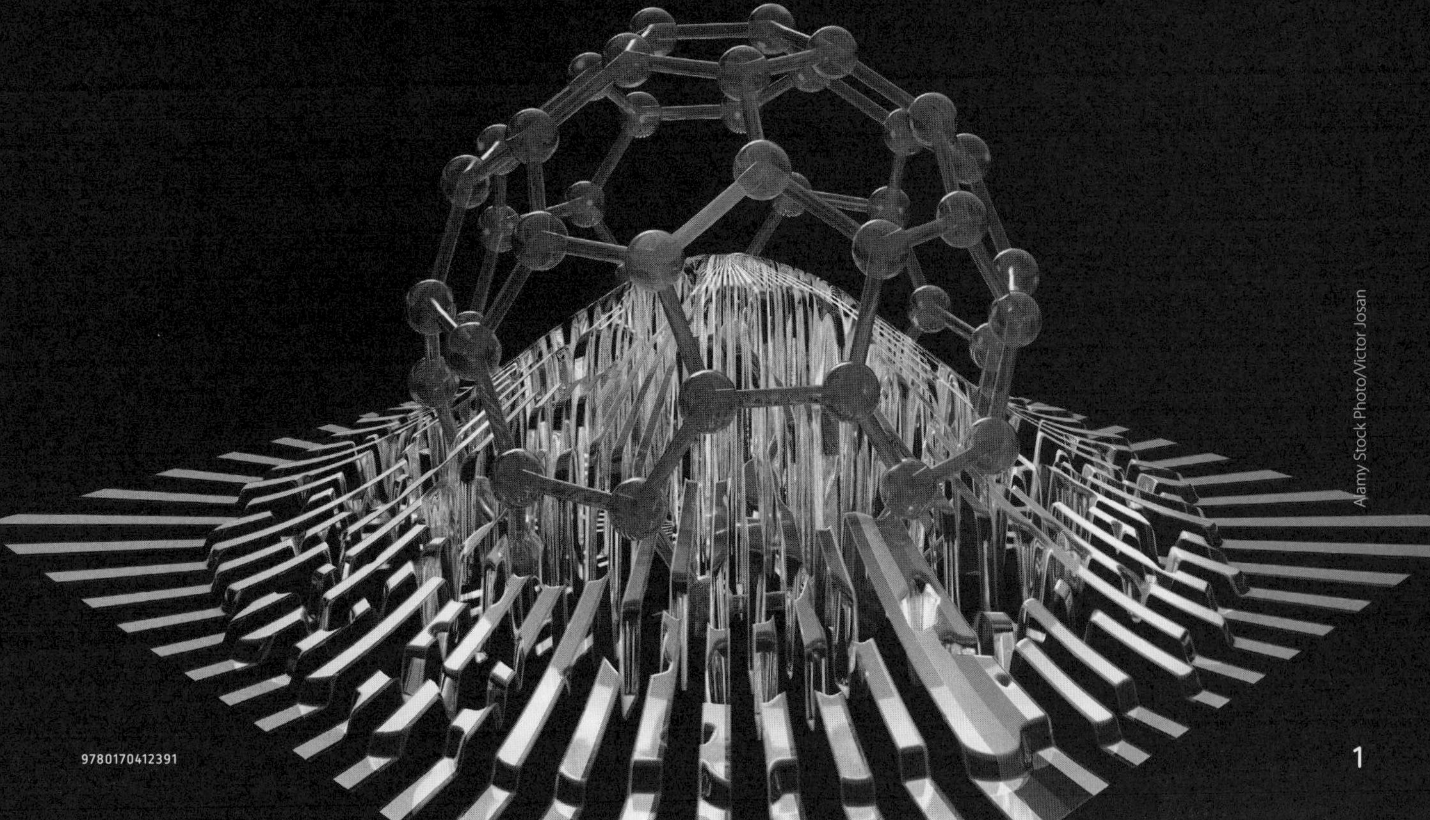

Alamy Stock Photo/Victor Josan

1 Periodic table and trends

Summary

- An element is a substance made from one type of atom. There are 118 known elements, each of which is listed on the periodic table of elements.
- The periodic table of elements lists the elements in order of atomic number, beginning with hydrogen (atomic number = 1) and currently finishing with oganesson (atomic number = 118), the largest known element.
- In the periodic table, the elements are arranged in periods (rows) and groups (columns). The groups are numbered from 1 to 18.
- Elements in the same group have very similar properties (e.g. in group 1 – lithium, sodium and potassium). Because of this, certain groups are given specific names. These include: group 1 – alkali metals; group 2 – alkaline earth metals; group 17 – halogens; and group 18 – noble gases.
- Patterns and trends in properties of the elements are evident from the way in which they are arranged in the periodic table. Because of this, the properties of an individual element can be predicted from its position.
- Elements can be classified as metals, non-metals or metalloids. Metals are shiny, conduct electricity in the solid state and are malleable. As we move down a group of the periodic table, the elements become more metallic. As we move across a period, the elements become less metallic. The non-metals are located on the right-hand side of the periodic table.
- Trends in the changing values of atomic radius, first ionisation energy and electronegativity can be observed. These trends can be explained by considering that as we move down a group, the number of electron shells increases, so that the attraction between the nucleus and the outermost electron in each element decreases. From left to right, across a period, the attraction between the nucleus and outermost electron increases due to the larger number of protons in the nucleus.
- Trends can also be observed in some of the compounds that elements form when reacted with oxygen, called oxides. Oxides formed from strongly metallic elements are alkaline when added to water, whereas those formed from strongly non-metallic elements are acidic. Oxides that can act as both acids and alkalis are known as 'amphoteric'.

9780170412391

1.1 The geography of the periodic table

The known elements can be classified as: halogens, noble gases, other non-metals, metalloids, alkali metals, alkaline earth metals, transition metals and rare earth metals.

1 Research the meaning of any terms from the list above with which you are unfamiliar.

2 Find out which elements belong to each class.

3 Colour code the periodic table (Figure 1.1) to show the location of each class of elements.

4 Indicate which class each colour represents in the box below the table.

5 Label groups 1 to 18 across the top of the periodic table.

6 Label periods 1 to 7 down the left side of the periodic table.

Key: Atomic number / Symbol / Element name

1 H hydrogen																	2 He helium
3 Li lithium	4 Be beryllium											5 B boron	6 C carbon	7 N nitrogen	8 O oxygen	9 F fluorine	10 Ne neon
11 Na sodium	12 Mg magnesium											13 Al aluminium	14 Si silicon	15 P phosphorus	16 S sulfur	17 Cl chlorine	18 Ar argon
19 K potassium	20 Ca calcium	21 Sc scandium	22 Ti titanium	23 V vanadium	24 Cr chromium	25 Mn manganese	26 Fe iron	27 Co cobalt	28 Ni nickel	29 Cu copper	30 Zn zinc	31 Ga gallium	32 Ge germanium	33 As arsenic	34 Se selenium	35 Br bromine	36 Kr krypton
37 Rb rubidium	38 Sr stronium	39 Y yttrium	40 Zr zirconium	41 Nb niobium	42 Mo molybdenum	43 Tc technetium	44 Ru ruthenium	45 Rh rhodium	46 Pd palladium	47 Ag silver	48 Cd cadmium	49 In indium	50 Sn tin	51 Sb antimony	52 Te tellurium	53 I iodine	54 Xe xenon
55 Cs caesium	56 Ba barium	57–71 lanthanides	72 Hf hafnium	73 Ta tantalum	74 W tungsten	75 Re rhenium	76 Os osmium	77 Ir iridium	78 Pt platinum	79 Au gold	80 Hg mercury	81 Tl thallium	82 Pb lead	83 Bi bismuth	84 Po polonium	85 At astatine	86 Rn radon
87 Fr francium	88 Ra radium	89–103 actinides	104 Rf rutherfordium	105 Db dubnium	106 Sg seaborgium	107 Bh bohrium	108 Hs hassium	109 Mt meitnerium	110 Ds darmstadtium	111 Rg roentgenium	112 Cn copernicium	113 Nh nihonium	114 Fl flerovium	115 Mc moscovium	116 Lv livermorium	117 Ts tennessine	118 Og oganesson

57 La lanthanium	58 Ce cerium	59 Pr praseodymium	60 Nd neodymium	61 Pm promethium	62 Sm samarium	63 Eu europium	64 Gd gadolinium	65 Tb terbium	66 Dy dysprosium	67 Ho holmium	68 Er erbium	69 Tm thulium	70 Yb ytterbium	71 Lu lutetium
89 Ac actinium	90 Th thorium	91 Pa protactinium	92 U uranium	93 Np neptunium	94 Pu plutonium	95 Am americium	96 Cm curium	97 Bk berkelium	98 Cf californium	99 Es einsteinium	100 Fm fermium	101 Md mendelevium	102 No nobelium	103 Lr lawrencium

FIGURE 1.1.1 The periodic table of elements

1.2 Important terms

Answer each of the clues below by writing the correct term on the following line. Create a mind map that includes all of the terms.

1 A substance that reacts with both an acid and a base. ____________________

2 A substance made from one type of atom. ____________________

3 A row in the periodic table ____________________

4 Half the distance between the nuclei of two neighbouring atoms is atomic ____________.

5 The region of the periodic table including groups 3–12 are ____________ metals.

6 A substance that has properties of both a metal and a non-metal. ____________________

7 A column in the periodic table. ____________________

8 The group 18 elements are the ________ gases.

9 The group 1 elements are the ____________ metals.

10 The name given to group 17 elements. ____________________

11 The ability of a bonded atom to attract electrons to itself. ____________________

12 The energy required to remove an electron from a neutral atom is the ______________ energy.

13 Summary or mind map of important terms.

1.3 Explaining trends in atomic radius

1 Source and draw representations of the atoms of beryllium, calcium and nitrogen. In your diagrams, ensure that the numbers of protons and neutrons are shown in the nucleus in the centre of the atom, and the arrangement of electrons is also shown.

a Beryllium and calcium are both in group 2 of the periodic table. Use your diagram to explain why the atomic radius of calcium is bigger than that of beryllium.

b Beryllium and nitrogen are both in period 2 of the periodic table. Use your diagram to explain why the atomic radius of nitrogen is smaller than that of beryllium. (Remember, protons are positively charged and electrons are negatively charged. Because their charges are opposite, they attract each other).

1.4 Explaining trends in electronegativity

Representations of the atoms of carbon, oxygen and silicon are shown in Figure 1.4.1. In these diagrams, the numbers of protons and neutrons are shown in the nucleus in the centre of the atom, and the arrangement of electrons is also shown.

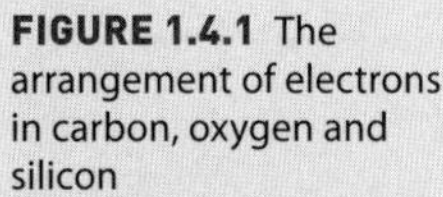

FIGURE 1.4.1 The arrangement of electrons in carbon, oxygen and silicon

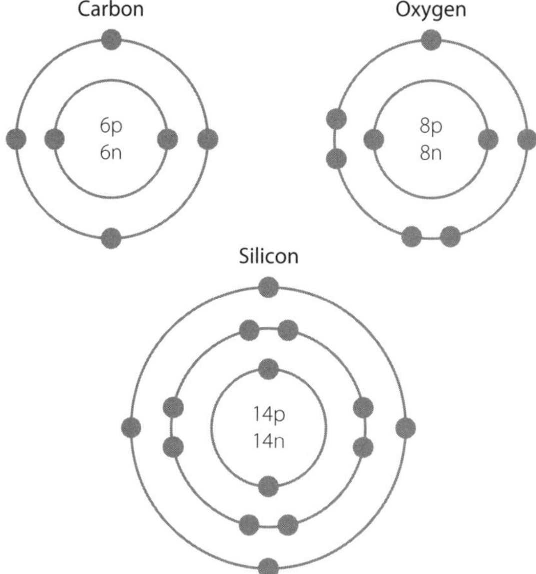

Electronegativity is the ability of a bonded atom to attract electrons to itself. Oxygen is the second most electronegative element in the periodic table, after fluorine. On the periodic table, electronegativity increases up the table and from left to right across the table.

1 Carbon and silicon are both in group 3 of the periodic table. Use Figure 1.4.1 to help explain why carbon is more electronegative than silicon.

____

____

____

____

____

2 Carbon and oxygen are in the same period. Use Figure 1.4.1 to explain why oxygen is more electronegative than carbon.

____

____

____

____

____

9780170412391

1.5 Explaining trends in first ionisation energy

The first ionisation energy is the energy required to remove an electron from a neutral atom. On the periodic table, the first ionisation energy of the elements increases up the table and from left to right.

1 Source and draw a representation of the atoms of boron, aluminium and fluorine. Make sure that the numbers of protons and neutrons are shown in the nucleus in the centre of the atom, and the arrangement of electrons is also shown.

2 Boron and aluminium are both in group 3 of the periodic table. Use your diagram to explain why boron has a greater first ionisation energy than aluminium.

3 Boron and fluorine are in the same period. Use your diagram to explain why fluorine has a greater first ionisation energy than boron.

1.6 Oxides

- An oxide is a compound of oxygen and another element.
- Metals form metal oxides and non-metals form non-metal oxides.
- Oxides can be classified as acidic, alkaline or amphoteric.
- The nature of the oxides of the elements across the periodic table from left to right changes from alkaline → amphoteric → acidic.

Answer the following questions. Circle to identify the correct answer and explain why the other answers are incorrect.

1 Which of the following reacts with both acid and alkaline solutions?

A Aluminium

B Sodium oxide

C Sulfur dioxide

D Aluminium oxide

2 Which element forms an oxide that reacts with water to produce an acidic solution?

A Methane

B Sodium

C Phosphorus

D Zinc

____

____

____

____

3 Which element burns in air to form an oxide which, when shaken with water, gives a solution with a pH greater than 7?

A Carbon

B Hydrogen

C Magnesium

D Sulfur

____

____

____

____

EVALUATION

1 Lithium and potassium are two elements found in group 1 of the periodic table. The elements in group 1 are known to be very reactive. Some data for lithium and potassium are provided in Table 1.7.

TABLE 1.7 Some properties of lithium and potassium

NAME	ATOMIC NUMBER	FIRST IONISATION ENERGY ($kJ\,mol^{-1}$)	ELECTRONEGATIVITY (PAULING SCALE)	ATOMIC RADIUS (pm)
Lithium	3	520	0.98	167
Potassium	19	419	0.82	280

a What is the commonly used name for the elements in group 1 to which potassium and lithium belong.

____

b Define the following terms.

i First ionisation energy

____

____

ii Electronegativity

____

____

c Suggest a reason why the atomic radius of lithium is smaller than that of potassium.

____

____

____

9780170412391

d Sodium is also a member of group 1, with an atomic number of 11. Suggest how the first ionisation energy of sodium would compare to that of lithium and potassium and explain your reasoning.

____

____

____

____

____

e Beryllium is element number 4 on the periodic table. Use the data in the table to predict what its electronegativity might be and explain your reasoning.

____

____

____

____

____

2 Atomic structure

Summary

- An atom is the smallest type of particle that can exist in a stable state.
- Atoms consist of a very small, but highly dense nucleus, where the vast majority of the mass of the atom is concentrated. The nucleus is surrounded by electrons, which are in 'orbit' around it.
- Atoms are made from three types of subatomic particles: protons, neutrons and electrons. Protons and neutrons have a virtually equal mass and exist in the nucleus of the atom. Protons are positively charged and neutrons are uncharged. Electrons have a negligible mass compared to that of a proton or neutron and are negatively charged.
- Atoms are held together by the strong nuclear force, which holds protons and neutrons together in the nucleus, and by the electrostatic attractive force, by which the negatively charged electrons are held in orbit around the positively charged nucleus.
- Each atom of an element has a unique symbol, and is represented by two numbers, the atomic number and the mass number.
- The atomic number is the number of protons, and the mass number is the total number of protons and neutrons in the nucleus. Hence, the number of neutrons in an atom can be determined by subtracting the atomic number from the mass number.
- For a neutral atom, the number of electrons is equal to the number of protons.
- Isotopes are two atoms of the same element, with the same number of protons, but different numbers of neutrons.
- The mass number of an atom is a whole number. However, on a periodic table, the relative atomic mass is written next to each element, which is not a whole number. The relative atomic mass is the average mass of an atom of the element, accounting for the different isotopes that are present.
- Electrons exist in energy levels or shells surrounding the nucleus. Electrons cannot exist in between these shells. Each shell can hold a maximum number of electrons: the first shell (closest to the nucleus) holds a maximum of 2 electrons; the second shell, 8 electrons; and the third shell, 18 electrons.
- The Aufbau principle states that the electron shells fill in order, lowest energy first.

9780170412391

- Electron shells can be divided into subshells, which contain groups of orbitals. Each orbital can hold two electrons. The number of different types of orbitals increases as the shell number increases. The first shell has an s orbital, the second shell has one s and three p orbitals, and the third shell has one s, three p and five d orbitals.
- The Pauli exclusion principle states that each electron has a unique location, so that no two electrons can occupy the same space. The electron configuration of each element can therefore be written in terms of the s, p and d notation.
- Hund's rule of maximum multiplicity indicates that each orbital within a subshell will be filled singly at first before the electrons are paired.

2.1 Important terms

1 Complete the following crossword using the clues on page 15.

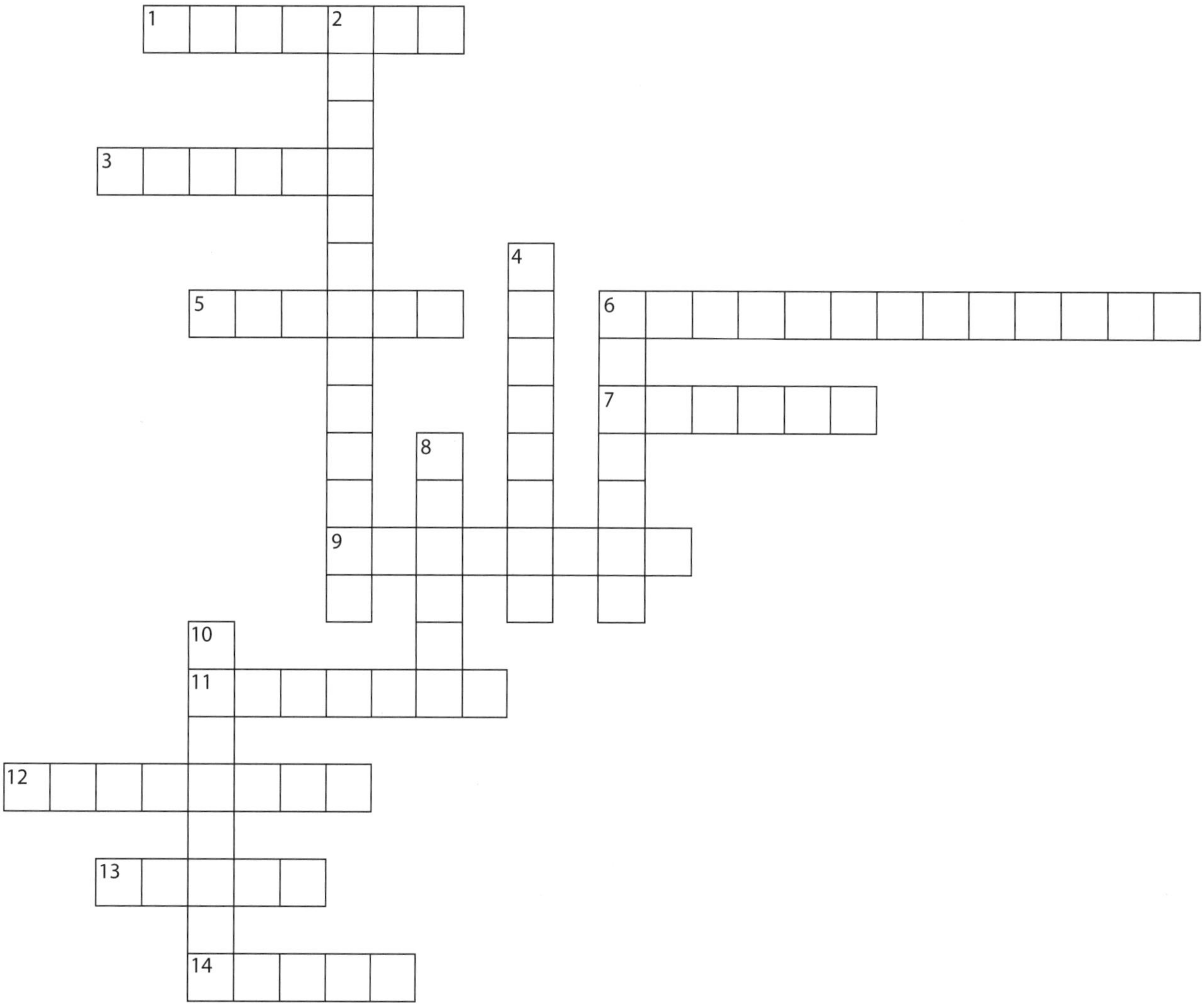

Across	Down
1 Dense central region of an atom **3** ________ number – number of protons in the nucleus of an atom **5** _________ force, which binds neutrons and protons together **6** Arrangement of electrons around the nucleus in energy levels **7** ________ principle – electrons fill shells in order, lowest energy first **9** Atoms of an element with varying numbers of neutrons **11** Substance made of atoms all with same number of protons **12** The s subshell contains only one ______________ **13** Particles of which all matter is made **14** Region around nucleus containing electrons with equal energy	**2** Type of attraction between oppositely charged objects **4** Negatively charged subatomic particles found outside the nucleus **6** Property of subatomic particles that may be positive or negative **8** Positively charged subatomic particle found in the nucleus **10** Uncharged subatomic particles found in the nucleus

2 Write a short summary about atomic structure using the terms in the crossword.

2.2 Numbers of protons, neutrons and electrons

1 Complete the following table.

SYMBOL	CHARGE	ATOMIC NUMBER	MASS NUMBER	NUMBER OF PROTONS	NUMBER OF NEUTRONS	NUMBER OF ELECTRONS
	0	3	7			
O	2–				9	
			24		12	10
				33	42	36
B	3+		7			
Kr					47	36
	1–			9	10	
	0	17	37			
		20	41			
	0		26			12
		29			33	27
	1+		63	29		

2.3 Electron configuration

1 What is the Aufbau principle?

____

____

____

____

2 Complete the sequence of subshells below, showing the order in which they are filled:

1s, 2s, 2p, ____, ____, ____, ____, 4p

3 Using s, p, d, f notation, write out the electron configurations of the first 36 elements next to the element symbols in the following table. Use shorthand notation based on the noble gas cores (e.g. Li = [He] $1s^1$). (Hint: chromium and copper are exceptions to the rule.)

PERIOD 1 ELEMENTS AND ELECTRON CONFIGURATION	PERIOD 2 ELEMENTS AND ELECTRON CONFIGURATION	PERIOD 3 ELEMENTS AND ELECTRON CONFIGURATION	PERIOD 4 ELEMENTS AND ELECTRON CONFIGURATION
H =	Li = [He] $1s^1$	Na =	K =
He =	Be =	Mg =	Ca =
	B =	Al =	Sc =
	C =	Si =	Ti =
	N =	P =	V =
	O =	S =	Cr =
	F =	Cl =	Mn =
	Ne =	Ar =	Fe =
			Co =
			Ni =
			Cu =
			Zn =
			Ga =
			Ge =
			As =
			Se =
			Br =
			Kr =

4 Use the table completed in question **3** to help you write the electron configurations for the following ions based on the notation for the noble gas core. (Remember that electrons are lost from the highest energy orbital first.)

a Hydrogen H^+ ______________________________

b Beryllium Be^{2+} ______________________________

c Nitrogen N^{3-} ______________________________

d Aluminium Al^{3+} ______________________________

e Sulfur S^{2-} ______________________________

f Potassium K^+ ______________________________

g Scandium Sc^{3+} ______________________________

h Vanadium V^{2+} ______________________________

i Manganese(II) Mn^{2+} ______________________________

j Manganese(III) Mn^{3+} ______________________________

k Nickel(II) Ni^{2+} ______________________________

l Copper(I) Cu^+ ______________________________

m Copper(II) Cu^{2+} ______________________________

n Germanium(IV) Ge^{4+} ______________________________

o Bromide Br^- ______________________________

2.4 Key ideas

1 Use the word list provided to fill in the blanks in the following sentences. (Hint: some words are used more than once.)

atoms	eighteen	exclusion	isotopes	maximum	nucleus	protons	speed
Aufbau	electrons	heavier	lighter	negative	orbitals	protons	symbol
blocks	elements	highest	lowest	neutrons	positive	shells	two
eight	equal	highest	mass	neutrons	protons	six	uncharged

a All matter is composed of extremely small particles called ______________.

b Atoms have an internal structure consisting of a tiny, dense central area called the ______________, surrounded largely by empty space.

c Atoms are formed out of three types of subatomic particles: ______________, ______________ and ______________.

d Protons and neutrons are ______________ subatomic particles found inside the nucleus.

e Electrons are ______________ subatomic particles found outside of the nucleus.

f Protons and neutrons, which are densely packed together in the nucleus, provide most of an atom's ______________.

g Electrons, which move around the nucleus at high ______________, occupy most of the volume of the atom but provide little of its mass.

h A proton has one ______________ charge and an electron has one ______________ charge, but a neutron is ______________.

i There are 92 different types of naturally occurring atoms. Each type is defined by the number of ______________ in its nucleus. For example, every carbon atom has six of these subatomic particles.

j Uncharged atoms have ______________ numbers of protons and electrons, so an uncharged carbon atom will have ______________ electrons.

k A substance consisting of only one type of atom is called an ______________. For example, the element carbon consists of atoms with six ______________ only.

l Each element is given a unique name and ______________ (e.g. magnesium, symbol Mg).

m Each type of atom can have varying numbers of ______________ in the nucleus. For example, carbon atoms can have 6, 7 or even 8 of them. These different forms of an atom are known as ______________.

n An atom's electrons have discrete amounts of energy and are located in specific energy levels, known as ______________ around the nucleus, depending on how much energy they have. Those in the innermost shells have the ______________ amount of energy.

o Each shell can accommodate up to a ______________ number of electrons. The first shell accommodates up to ______________, the second shell ____________ and the third shell up to ______________ each.

p Each shell contains a series of ______________, which can be labelled s, p, d or f.

q The ____________ principle states that the electron shells are filled, lowest in energy first.

r The Pauli ______________ principle states that each electron has a unique location within the atom, which can be represented using the s, p, d, f notation.

s When ions are formed, electrons are lost from the ______________ energy level first.

t The periodic table can be divided into __________, according to the subshell in which the outermost electron exists.

9780170412391

1 Our contemporary model of the atom includes three types of subatomic particles, the proton, neutron and electron.

a Write a paragraph to compare and contrast the proton, neutron and electron, by referring to their mass, their charge and their location within the atom.

b Two forces, electrostatic attraction and the strong nuclear force are essential in keeping the nucleus stable. Explain why each force is so important.

i Electrostatic force

ii Strong nuclear force

2 The element symbols, atomic numbers and mass numbers of two species of unknown elements are shown as follows.

$$^{49}_{22}X \text{ and } ^{42}_{20}Z^{2+}$$

a Complete the following table for the two species X and Z.

SPECIES	NUMBER OF PROTONS	NUMBER OF NEUTRONS	NUMBER OF ELECTRONS
$^{49}_{22}X$			
$^{42}_{20}Z^{2+}$			

b Write the electron configuration for $^{49}_{22}X$ using s, p, d, f notation.

____

____

c Explain why element Z has a charge of 2+.

____

____

____

____

____

3 Chromium has an electron configuration that does not exactly fit the pattern shown by the first row of transition metal elements.

a Using shorthand notation based on the noble gas core, write the electron configuration of titanium, vanadium and chromium.

i Ti ______________________________

ii V ______________________________

iii Cr ______________________________

b Explain why the electron configuration of chromium is unusual.

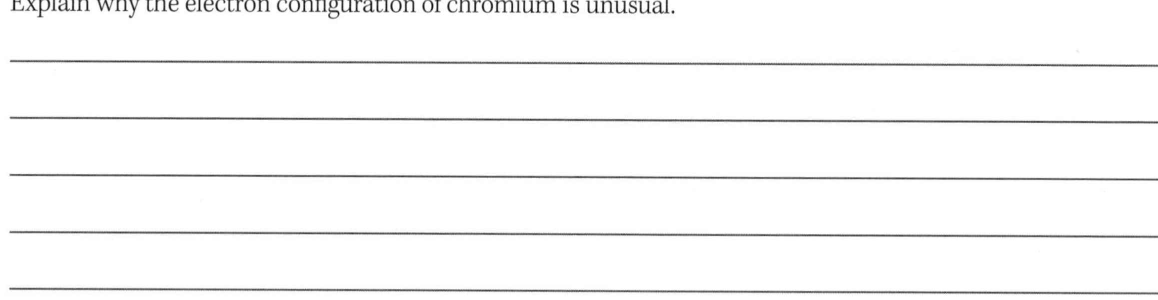

9780170412391

3 Introduction to bonding

LEARNING

Summary

- A compound is formed when two or more elements are combined in a fixed ratio to form a new substance that has different chemical properties from its constituent elements.
- Elements form chemical bonds with each other to achieve a more stable electron configuration, which is often the same as that of a noble gas atom. Therefore, the types of bonds and substances that an element will form are determined by its electron configuration.
- Metallic bonds occur between atoms of metallic elements, where the valence electrons from each atom become 'delocalised', moving randomly throughout the lattice of positively charged ions that remain. The electrostatic force of attraction between the positively charged metal ions and the negatively charged delocalised electrons holds the structure together. Metallic bonds do not result in the formation of a compound, but in an alloy, where the resulting mixture shares properties of the constituent elements.
- Ionic bonding occurs between a metal and a non-metal.
    - Metal atoms lose their valence electrons, forming a positive ion (a cation). Non-metals can gain electrons, so that they achieve a noble gas configuration, and so form a negatively charged ion (an anion). There is then an electrostatic attraction between the positive and negative ions, forming a compound. The ions combine in a specific ratio, such that the total number of electrons lost by the positive ions is equal to that gained by the negative ions.
    - Ionic bonds are non-directional: every positive ion will attract every negative ion. Hence ionic compounds form lattices of regularly repeating arrangements of ions
    - The charge on an ion can be predicted from the position of the element in the periodic table.
    - Transition elements can form more than one different ion, depending on the circumstances, and so in transition metal compounds, roman numerals are used to indicate the charge on the metal ion.
- Covalent bonding occurs between non-metallic elements.
    - The non-metallic elements share electrons between overlapping shells, so that the valence shells of both elements are filled.
    - A covalent bond is a shared pair of electrons. Pairs of valence electrons which are not shared between atoms are known as non-bonding, or lone pairs of electrons.
    - If more than one pair of electrons is shared between two atoms, then multiple bonds can be formed.
    - Lewis diagrams can be used to represent covalent molecules, using lines to represent a pair of electrons.

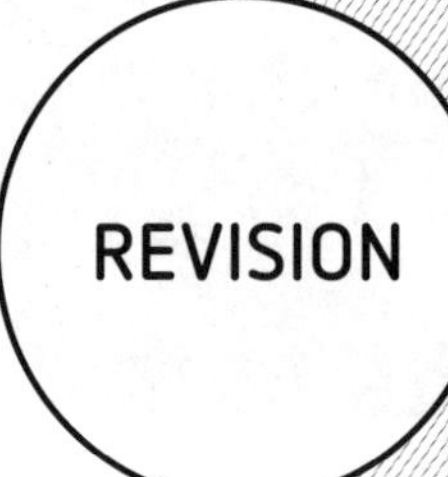

3.1 Important terms

1 Complete the following crossword using the clues on page 23.

Across	Down
2 A pair of valence electrons not involved in the covalent bond	**1** Negatively charged atom or group of atoms
4 A pure substance composed of more than one type of element bonded together	**3** Type of bond involving attraction between cations and anions in a compound
9 Positively charged atom or group of atoms	**5** Type of detached electrons that are free to move about within the structure
11 Type of dot formula used to represent valence electrons of atoms or ions	**6** Type of bond involving shared electron pairs that holds atoms together within a molecule
13 The number of single bonds with hydrogen atoms that the atom could form	**7** Type of bonding involving attraction between delocalised valence electrons and metal cations
14 Charged atom or group of atoms	**8** Type of covalent bond in which two pairs of electrons are shared
	10 Type of electrons found in the outer shell of an atom in its ground state
	12 Type of attraction between shared electrons and the nuclei of the atoms involved

2 Write a short summary about bonding using the words in the crossword.

3.2 Comparison of bonding types

In the following table, describe the distinctive features of each type of bonding. Then identify features common to all four types of bonding.

BONDING MODELS	DISTINCTIVE FEATURES	COMMON FEATURES
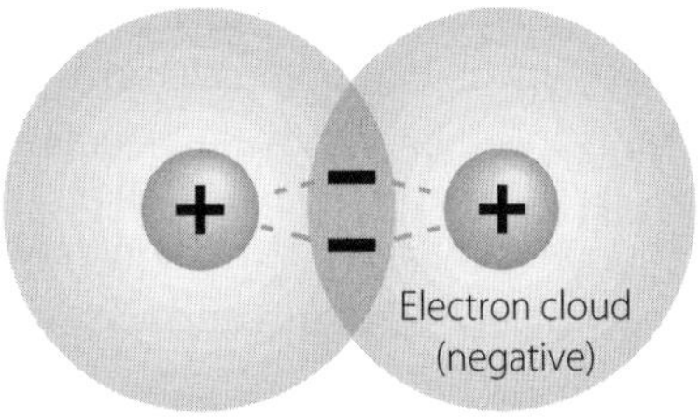 + + Electron cloud (negative)		
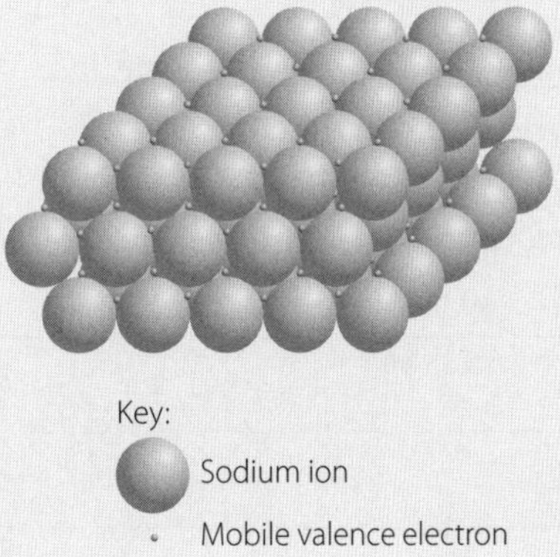 Key: Sodium ion Mobile valence electron		
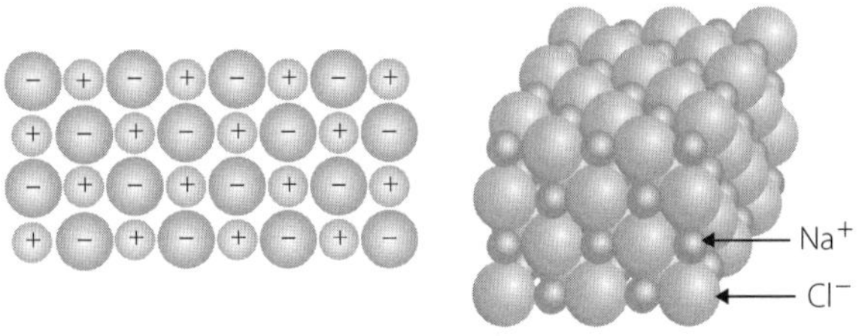 Na^+ Cl^-		

9780170412391

3.3 Key ideas

1 Match each term with the appropriate description using an arrow.

Term		Description
a Anion		**A** Attraction between cations and anions in an ionic compound
b Bonding pair		**B** Attraction between delocalised valence electrons and metal cations
c Cation		**C** Attraction between protons and electrons of participating atoms
d Chemical bond		**D** Attraction between shared electrons and nuclei of atoms involved
e Compound		**E** Bonding that occurs in most directions
f Covalent bonding		**F** Charged atom or group of atoms
g Molecular compound		**G** Compound that is made up of positive ions and negative ions
h Delocalised electrons		**H** Covalent bond with two pairs of electrons are shared
i Directional bonding		**I** Covalent bond with more than one pair of shared electrons
j Double covalent bond		**J** Detached electrons that are free to move about within the metal
k Ion		**K** Direct line bonding between adjacent atoms in a molecule
l Ionic bonding		**L** Discrete molecule in which atoms are joined by covalent bonding
m Ionic compound		**M** Electrons in the outer shell of an atom in its ground state
n Metallic bonding		**N** Ions that consist of two or more atoms strongly covalently bonded
o Multiple covalent bond		**O** Negatively charged atom or group of atoms
p Non-bonding (lone) pair		**P** Pair of electrons shared by two atoms resulting in covalent bonding
q Non-directional bonding		**Q** Pair of valence electrons that are not involved in the covalent bond
r Polyatomic ion		**R** Positively charged atom or group of atoms
s Valence electrons		**S** Pure substance composed of different atoms bonded together
t Valency		**T** Number of single bonds with a H atom that the atom could form

3.4 Ionic and covalent formulas

TABLE 3.4.1 Examples of common polyatomic ions and the ionic compounds made from them

NAME OF ION	FORMULA	VALENCY	EXAMPLE OF A COMPOUND
Ammonium	NH_4^+	+1	Ammonium chloride
Hydroxide	OH^-	−1	Iron(III) hydroxide
Nitrate	NO_3^-	−1	Silver nitrate
Sulfate	SO_4^{2-}	−2	Copper(II) sulfate
Carbonate	CO_3^{2-}	−2	Calcium carbonate
Phosphate	PO_4^{3-}	−3	Sodium phosphate

1 Use Table 3.4.1 and the periodic table (Chapter 1, page 3) to work out the formulas of the following ionic compounds.

a Sodium hydroxide

b Calcium nitrate

c Zinc(II) sulfate

d Cobalt(III) oxide

e Aluminium carbonate

f Copper(I) phosphate

2 Write the formulas of the following covalent compounds.

a Silicon tetrafluoride

b Carbon dioxide

c Carbon monoxide

d Selenium dichloride

e Dichlorine oxide

f Nitrogen tribromide

1 Complete the table to show the molecular formula and valence structure for each molecule.

NAME	MOLECULAR FORMULA	VALENCE STRUCTURE
Carbon tetrachloride		
	CO_2	
		Cl–B(–Cl)–Cl (B bonded to three Cl)
Ammonia	NH_3	

2 Write formulas for the following compounds and state whether the bonding within them is ionic or covalent.

NAME	FORMULA	BONDING TYPE
Sodium oxide		
Sulfur dioxide		
Boron sulfide		
Manganese(II) chloride		
Selenium iodide		

3 When magnesium is reacted with oxygen, a chemical reaction occurs producing magnesium oxide (MgO), which has completely different properties from that of its constituent elements. However, when magnesium is mixed with sodium, the resulting substance is still metallic and has properties of both magnesium and sodium.

Write a paragraph to explain:

a what occurs when magnesium reacts with oxygen to form magnesium oxide.

b how the bonding differs in both products.

4 Phosphorus trichloride is described as having a 'lone pair' of electrons within its molecular structure. Explain, using an appropriate diagram, what is meant by a lone pair of electrons.

9780170412391

4 Isotopes

LEARNING

Summary

- Isotopes are two atoms of the same element, but with different masses, due to their different numbers of neutrons in the nucleus.
- Isotopes are represented using the notation ^{A}X or X-*A*, where *A* is the mass number and X is the elemental symbol.
- In any naturally occurring sample of an element there is a range of isotopes present, and the abundance of each isotope is constant, regardless of the origin of the sample.
- The relative atomic mass (RAM) of an element is the average mass of one atom of an element, which accounts for the different isotopes that are present and their relative abundance.
- Isotopes have identical chemical properties due to their identical electron configuration; however, they have different physical properties because of the differing numbers of particles in the nucleus. As such, isotopes are used for many applications within society.
- Some isotopes possess the physical property of radioactivity. This occurs when the forces within the nucleus are imbalanced as a result of the specific number of protons and neutrons present and the nucleus becomes unstable. The nucleus decays, releasing high energy particles or radiation of varying forms. The half-life, amount of time taken for half of the nuclei present in a sample to decay, is constant. The half-life can be measured and used to calculate the rate of the radioactive decay of an isotope.

REVISION

4.1 Smoke detectors

FIGURE 4.1.1 A smoke detector

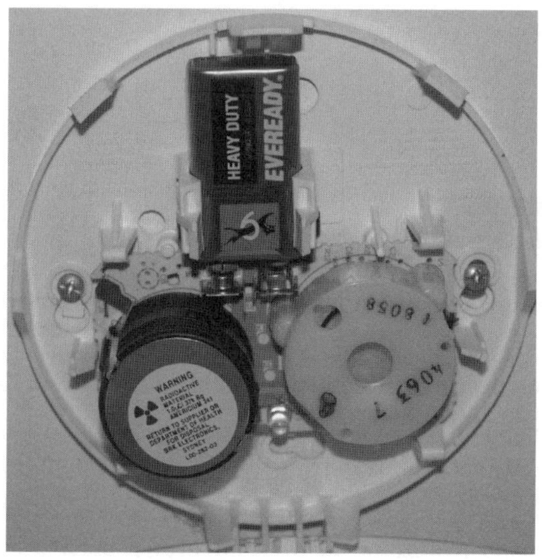

Figure 4.1.1 shows the interior of a smoke alarm or detector. It consists of an air-filled ionisation chamber (lower right) and a radioactive source of americium-241 (lower left). Americium (Am) is not a naturally occurring element; rather, it is a by-product of reactions occurring inside nuclear reactors.

Americium-241 is an unstable radioactive isotope and its nucleus will spontaneously eject alpha particles.

The alpha particles emitted by the americium isotope bombard the air inside the ionisation chamber, causing the air molecules to become electrically charged. This allows a small electric current to flow in the electrical field surrounding the chamber.

If smoke particles pass through the chamber, they also become ionised by alpha particles but because their mass is greater than that of air molecules, their acceleration in the electrical field is slower. This reduces the ionisation current, triggering a piezoelectric sounder, which emits a high-pitched squeal of about 85 decibels. (Sounds of about 75 decibels are needed to wake a sound sleeper.)

As the half-life of the americium isotope is about 433 years, it will emit alpha particles well beyond the 10-year lifespan of the detector. When the americium atoms decay, they form an isotope of the element neptunium (Np), which is much more stable with a half-life of 2.5 million years.

The ionising radiation of emitted alpha particles travels only a few centimetres in the air before the particles are neutralised, so they are unlikely to encounter human body cells.

FIGURE 4.1.2 A radioactive atom of americium-241 will decay to neptunium-237, emitting an ionising alpha particle.

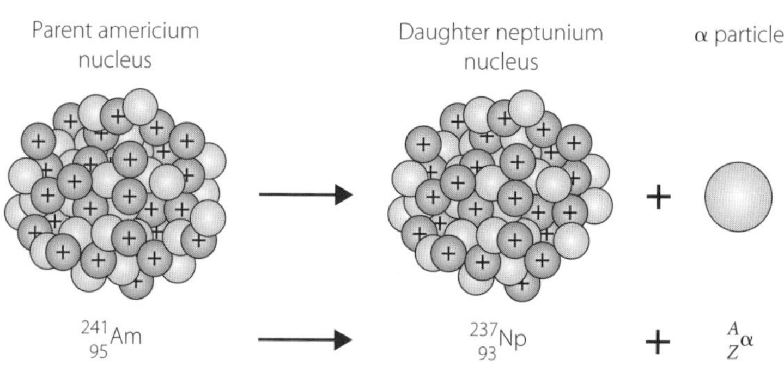

1 Describe the radioactive decay that occurs in a smoke detector.

2 What does the sign on the lower left chamber of the smoke detector in Figure 4.1.1 mean?

3 How many protons and neutrons does $^{241}_{95}Am$ have? How many does $^{237}_{93}Np$ have?

4 Which nuclear particles must an alpha particle consist of? How do you know?

5 What do alpha particles do to air and smoke molecules when they encounter them?

6 Why is a battery required in a smoke detector?

7 How can an electrical current flow through the ionisation chamber?

______

______

______

8 Why do ionised smoke particles reduce the ionisation current?

______

______

______

9 If a smoke detector starts making a repeated beeping sound, what would that mean?

______

______

______

10 Why is the alpha particle radiation emitted by a smoke detector not harmful to humans?

______

______

______

4.2 Isotope or not?

The following table provides some data about the number of subatomic particles in a series of different atoms.

1 Complete the table to provide data for each the atoms shown.

ATOM	ATOMIC NUMBER	MASS NUMBER	NUMBER OF PROTONS	NUMBER OF NEUTRONS	NUMBER OF ELECTRONS
Z	7	16		9	
Y				16	12
X	18	34			
W		18		11	
V		25	11	14	
U		15			7
T		22		11	

9780170412391

2 Which atoms in question **1** are isotopes of each other?

4.3 Relative atomic mass calculations

Tin has 10 isotopes, as shown in Table 4.3.1.

TABLE 4.3.1 The percentage abundance of the 10 isotopes of tin

MASS NUMBER	PERCENTAGE ABUNDANCE (%)
112	1.0
114	0.7
115	0.3
116	14.5
117	7.7
118	24.2
119	8.6
120	32.6
122	4.6
124	5.8

1 Calculate the RAM of tin.

2 Element R has two isotopes, with mass numbers of 30 and 33. If the RAM of R is 30.85, calculate the percentage abundance of each of the isotopes.

1 Define the term 'isotope'.

2 Define the 'RAM' of an element.

3 The RAM of element Y is 33.15. It has two isotopes: one with a mass number of 33 and the other with a mass number of 34. Calculate the percentage abundances of each isotope.

4 The process of photosynthesis is a series of complex biochemical reactions, which occur in the presence of sunlight and chlorophyll, a green pigment found in leaves and other plant structures. Photosynthesis can be summarised by the equation:

$$6CO_2 + 6H_2O \rightarrow C_6H_{12}O_6 + 6O_2$$

$$\text{Carbon dioxide} + \text{water} \rightarrow \text{glucose} + \text{oxygen}$$

Our understanding of photosynthesis was developed partially through the use of radioisotope labelling. Plants were grown in an environment containing carbon dioxide made with the radioactive carbon-14 isotope rather than the more common carbon-12. Subsequent analysis of the plants revealed that high levels of radioactive carbon-14 were found in plant-made glucose molecules.

a Explain why carbon-12 and carbon-14 have the same chemical properties, yet different physical properties.

______

______

______

______

______

b Explain how the properties of isotopes described in your answer to part **a** were useful in the study of photosynthesis.

______

______

______

______

______

5 Analytical techniques

LEARNING

Summary

- Information about the identity, purity and quantity of different chemicals can be obtained by using a variety of different machines.
- Qualitative techniques determine the identity of a substance being analysed.
- Quantitative techniques determine the amount of substance present in a sample.
- Mass spectrometry:
  - Can distinguish between different masses of elements or compounds in a sample – it ionises the particles and then separates them according to their mass.
  - Can be used to determine the isotopic composition of an element, and hence the relative atomic mass of an element.
- Absorption spectroscopy:
  - When energy such as light or heat is shone at a sample of atoms, specific frequencies of this energy are absorbed by the atoms, causing electrons to move (transition) to higher energy levels. Because the electron configuration of the atoms of each element is unique, the pattern of absorbed frequencies (the absorption spectrum) is also unique to each element. Hence, the presence of an element within a sample can be determined by analysis of the frequencies of radiation absorbed by the sample.
  - The amount of energy absorbed by an element is directly proportional to the amount of that element within a sample. Absorption spectroscopy can be used quantitatively; that is, to determine the amount of substance present.
- Emission spectroscopy:
  - Once atoms have absorbed energy, after a short time, that energy is released, as the electrons in higher energy levels than normal (excited state) fall back to their original levels (ground state). The frequencies of light that are released are therefore also unique to that element, again because of the unique electron configuration of its atoms. Analysis of this emitted light is called emission spectroscopy and can also be used to determine the identity of the element. The simplest analysis is a visual inspection of the colour of light produced in a flame – the flame test.
- Atomic absorption spectroscopy:
  - Can be used to determine the concentration of a specific metal ion within a sample. The radiation used is emitted from a lamp made from the specific metal being tested and will measure only the concentration of the intended metal ions.
  - A series of known standards is prepared and the concentration of each is measured using the spectrometer. Hence, a calibration curve of absorption versus concentration can be drawn.
  - When the absorption of the sample is measured, the unknown concentration can be interpolated from the calibration curve.

REVISION

5.1 Important terms

1 Complete the crossword on analytical techniques.

Across	Down
1 A type of analysis that measures the amount of a substance in a sample	**1** A type of analysis that determines the identity of a substance
2 Abbreviation for atomic absorption spectroscopy	**3** A type of spectroscopy that measures the extent that radiation is not transmitted by a sample
4 A particle of light	**5** A type of spectroscopy that measures the radiation produced by a sample
8 Mass ___________ – a technique to separate a mixture of sample according to their mass	**6** The scientist who linked the emission of discrete wavelengths of light to electron transitions
	7 _______ test – a type of emission spectroscopy carried out using the naked eye

9780170412391

2 Use the answers from the crossword to write a short summary of analytical techniques.

5.2 Constructing and using calibration curves

CSIRO/Scienceimage

FIGURE 5.2.1 Dr Alan Walsh, inventor of atomic absorption spectroscopy

In the 1950s, atomic absorption spectroscopy (AAS) was developed by CSIRO scientist Dr Alan Walsh. It is an analytical technique used to determine the unknown concentration of an element based on the amount of light it absorbs compared to samples of known concentration.

1 Use the data in Table 5.2.1 to construct a calibration curve on the grid for known concentrations of silver.

TABLE 5.2.1 The absorbance of silver at different known concentrations

SILVER CONCENTRATION (ppm)	ABSORBANCE
0.00	0.00
1.00	0.17
2.00	0.34
3.00	0.48
4.00	0.65
5.00	0.83
6.00	1.01

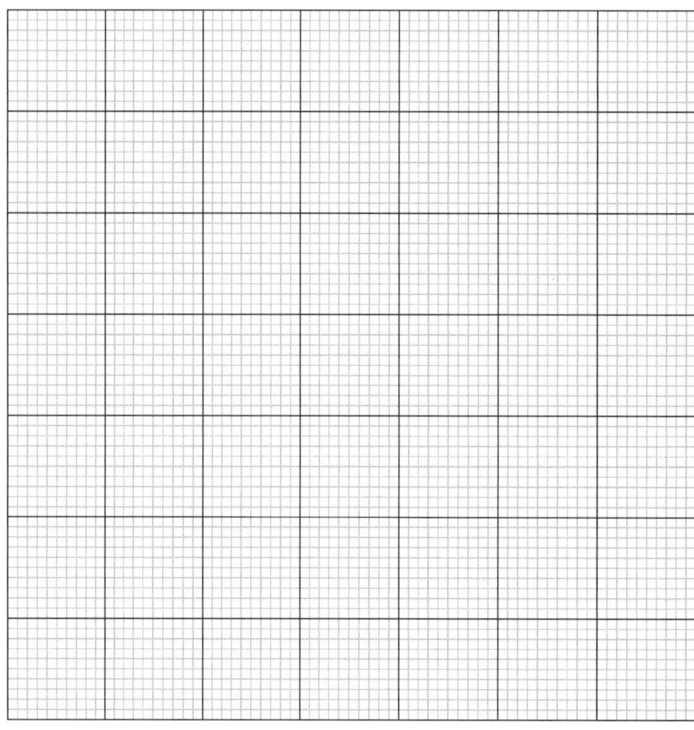

2 Use the curve to determine the concentration of silver in a sample with an absorbance of 0.53.

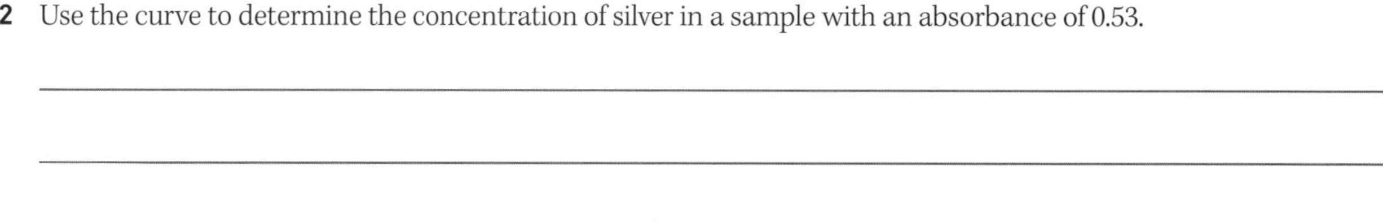

3 Use the data in Table 5.2.2 to construct a calibration curve on the grid for known concentrations of cadmium.

TABLE 5.2.2 The absorbance of cadmium at different known concentrations

CADMIUM CONCENTRATION($mg.L^{-1}$)	ABSORBANCE
0.00	0.000
1.00	0.038
2.00	0.082
3.00	0.120
4.00	0.160
5.00	0.200
6.00	0.240

9780170412391

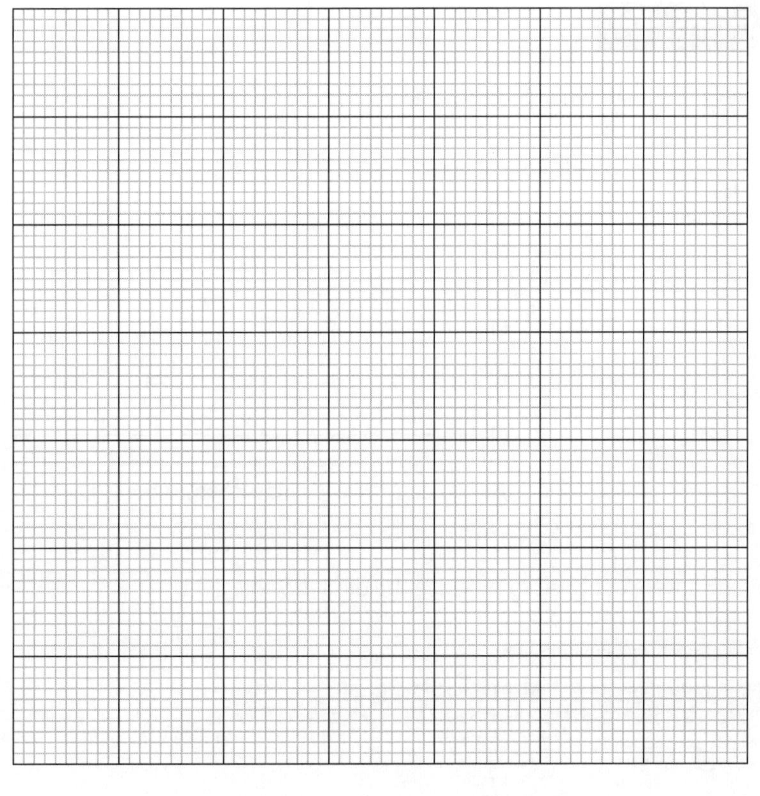

4 Use the curve to determine the concentration of cadmium in a sample with an absorbance of 0.190.

5 The development of AAS by Dr Walsh was a significant scientific discovery. AAS enabled very small concentrations of metal ions to be accurately and selectively determined. Explain how the atomic absorption spectrometer could be adjusted to test for the presence of either silver or cadmium in Questions **1-4**.

5.3 Mass spectrometry

The mass spectrum of magnesium is shown in Figure 5.3.1.

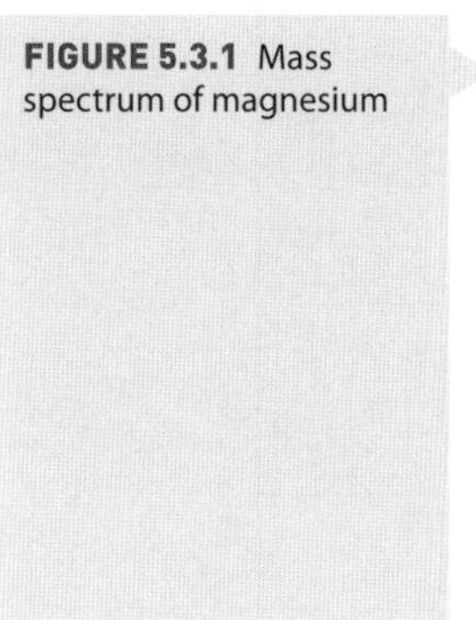

FIGURE 5.3.1 Mass spectrum of magnesium

The data from the spectrum were analysed to produce the data in Table 5.3.1.

TABLE 5.3.1 Data derived from the spectrum of magnesium

MASS:CHARGE RATIO	PERCENTAGE ABUNDANCE
24	78.70
25	10.13
26	11.17

1 What information does the mass spectrum provide about the isotopic composition of magnesium?

2 Use the data to calculate the RAM of magnesium.

9780170412391

5.4 Emission and absorption spectroscopy

Figure 5.4.1 shows the emission spectra of hydrogen atoms (left) and sodium atoms (right).

Hydrogen

Sodium

Roland Smith

FIGURE 5.4.1 Emission spectra for the elements hydrogen and sodium

Absorption spectra

1 Describe the difference between the ground and excited states of an atom.

2 Explain what happens when an electron absorbs an appropriate amount of energy.

3 What do each of the lines shown on the spectra in Figure 5.4.1 represent?

4 Why are the spectra of the two elements different?

5 Underneath each of the images above, draw a sketch of what the absorption spectra for hydrogen and sodium would look like. Explain your reasoning.

9780170412391

1 Lead poisoning is a disease in humans caused by excessively high exposure to lead (Pb). Due to safety concerns, the lead content of commercially traded fish that is intended for consumption by humans should not exceed 0.5 milligrams of lead per kilogram of fish consumed. One way to determine the lead content in fish is to analyse samples using AAS, as follows.

A mass of 0.1373 g of freeze-dried fish tissue is dissolved in 2.00 mL of nitric acid to dissolve all of the lead present. The 2 mL of solution is then diluted to 500 mL with water to form Solution A. Then 1.00 mL of Solution A is further diluted with water to 250 mL to form Solution B. AAS is used to measure the absorbance of Solution B, which is found to be 1.03. Next, the absorbance of a series of lead solutions of known concentrations was measured using AAS. The results obtained are shown in Table 5.5.1.

TABLE 5.5.1 The absorbance of a series of lead solutions of known concentration

CONCENTRATION OF PB ($\mu g.L^{-1}$)	ABSORBANCE
0.1	0.3
0.2	0.6
0.3	0.9
0.4	1.2
0.5	1.5

a Use the data in Table 5.5.1 to plot a calibration curve of absorbance against concentration and use the graph to find the concentration of lead in Solution B in $\mu g.L^{-1}$ of water.

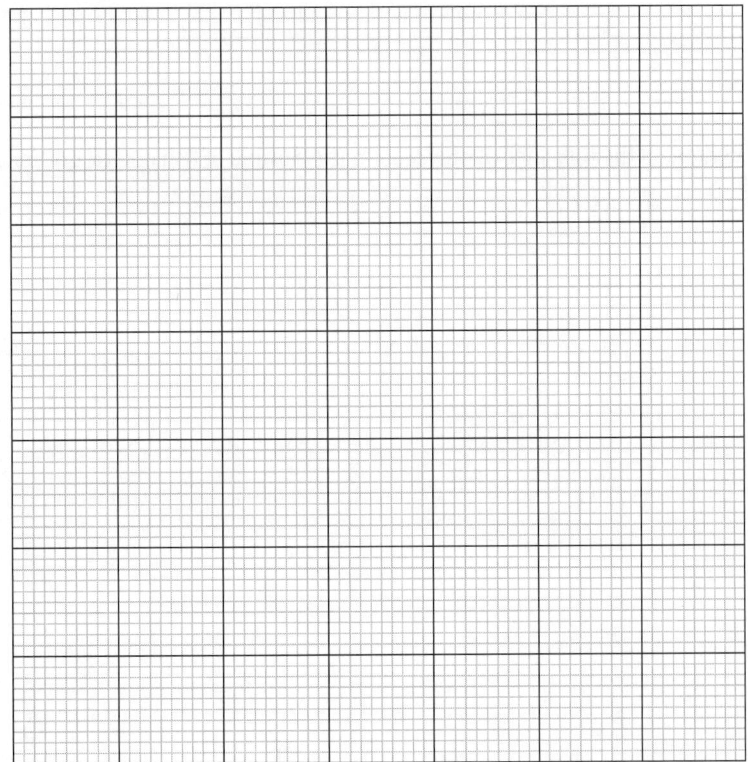

Therefore, the concentration of lead in Solution B is:________ μg.L^{-1}.

b Use your result to find the concentration of lead in Solution A in μg.L^{-1}.

c If we assume that the lead was dissolved in 500 mL of water to form Solution A, find the mass of lead in milligrams that must have been present in the original fish sample. (Note that 1 mg = 1000 μg).

d Should this fish sample be sold for human consumption? Justify your answer.

e Explain how AAS detects the presence of only lead in the sample of fish and not other elements.

9780170412391

6 Compounds and mixtures

LEARNING

Summary

- In Chapter 6, the following classifications were made:
  - materials as homogeneous or heterogeneous
  - materials as pure or impure
  - pure substances as elements or compounds
  - properties as physical or chemical
  - microscopes as optical or electron
  - liquids as miscible or immiscible
  - solutes as soluble or insoluble in different solvents
  - alloys as substitution or interstitial.
- In Chapter 6, the following observations were associated with a chemical reaction occurring:
  - solid forming or disappearing
  - gas evolving
  - colour changing
  - temperature changing.
- In Chapter 6, the following processes were associated with separation of liquids:
  - filtration
  - distillation
  - use of separatory funnels
  - vaporisation
  - extraction.
- In Chapter 6, the following substances were associated with nanotechnology:
  - fullerenes
  - graphene
  - carbon nanotubes
  - gold nanoparticles.

REVISION

6.1 Important terms

1 Complete the following statements by filling in the missing words from the word list. (Hint: some words are used more than once).

Word list

boiling	evolution	heating	natural	precipitation	synthetic
chemical	flammability	liquid	physical	solid	temperature
desalination	gas	melting	precipitation	solubility	uniformity
evolution	heat	microscope			

a The ________________ state of matter has a defined volume and shape.

b The ________________ state of matter has a defined volume, but not a defined shape.

c The ________________ state of matter has neither a defined volume, nor a defined shape.

d Substances that are found in nature are called ________________.

e Substances that are created in a laboratory are called ________________.

f A ________________ process starts and finishes with the same substances.

g A ________________ process starts and finishes with different substances.

h A ________________ enables scientists to view phenomena too small to be seen with the naked eye.

i ________________ converts a substance in solid form into liquid form.

j ________________ converts a substance in liquid form into gas form.

k ________________ is a measure of a solute's ability to dissolve in a solvent.

l ________________ is a measure of a substance's ability to burn.

m In meteorology ________________ means rain.

n In chemistry ________________ means a solid is formed.

o In biology ________________ means a change in heritable characteristics over successive generations.

p In chemistry ________________ means a gas is produced.

q ________________ is a quantitative measure of a substance's hotness or coldness.

r ________________ is increasing the temperature of a substance.

s ______________ energy is a function of temperature, mass and the type of substance.

t ________________ converts salt water into fresh water.

u ________________ is a measure of a substance's sameness of composition throughout.

2 Draw an arrow to match each term with its description (or list letter pairs that go together).

Term	Description
a Amorphous	**A** Composed of only one substance
b Carbon nanotube	**B** Composed of only one type of atom
c Colloid	**C** Can be broken down into simpler substances by chemical processes
d Compounds	**D** The substance that is dissolved in a solution
e Distillate	**E** The substance in which the solute of a solutions is dissolved into
f Distillation	**F** A technique for separating the solvent from a solution, when the solvent is required to be kept
g Elements	**G** The solvent kept after distillation
h Fullerene	**H** Liquids that do not mix
i Graphene	**I** A technique for separating the solvent from a solution, when the solute is required to be kept
j Heterogeneous	**J** A mixture of uniform composition
k Homogeneous	**K** A mixture of non-uniform composition
l Immiscible	**L** A non-crystalline solid
m Interstitial	**M** An alloy where the dopant has been substituted in the crystalline structure with the main atom
n Nanomaterial	**N** An alloy where the dopant occupies the space between main atoms
o Nanotechnology	**O** A branch of science dealing with particles in the range 1–100 nm
p Pure substances	**P** A substance made up of, or incorporates, particles in the range 1–100 nm
q Solutes	**Q** A nearly spherical arrangement of 20–84 covalently bound carbon atoms
r Solvent	**R** A sheet of covalently bound carbon atoms
s Substitution	**S** A rolled up sheet of graphene capped with fullerene(s)
t Vaporisation	**T** A mixture in which tiny clusters of particles are dispersed through another substance

3 Explain the differences between the terms in each of the following word pairs.

a Diesel and petrol

____

____

b Hydrophilic and hydrophobic

____

____

c Natural and synthetic diamonds

____

____

d The n-type and p-type regions of a solar panel

____

____

4 Use the clues on page 51 to complete the crossword. Some clues have two-word answers, but do not insert a space between the two words of your answer. Some answers are chemical formulas; for example, if the answer was H_2O, write this as 'HTWOO' into the crossword.

9780170412391

Across	Down
1 Observations made with the naked eye can discern objects at the millimetre scale	**1** Makes up 95% of natural gas
5 The substance into which the solute of a solutions is dissolved	**2** Chemical formula for methane
6 Potassium hydroxide	**3** Instrument that can distinguish individual bacterium at the micrometre scale
7 Can be broken down into simpler substances by chemical processes	**4** Fuel of choice in mobile situations
9 Gas in which combustion occurs (chemical formula)	**5** Substance that is dissolved in a solution
10 Gas used to fill airships today	**8** British chemist who isolated potassium
14 Instrument that has a resolution of < 0.1 nm	**11** Two physical properties are electrical and thermal ________
16 Type of technology dealing with particles in the range of 1–100 nm	**12** Calcium oxide
17 Type of technology used to isolate potassium	**13** Chemical formula of table salt
	15 Gas produced from combustion of fossil fuels (chemical formula)
	16 Toxic gas produced by reaction of copper and nitric acid (chemical formula)

5 Use the answers from the crossword to write a summary about compounds and mixtures.

6.2 Substances

1 Complete the following table, comparing the properties of substances.

SUBSTANCE	PURITY	HOMOGENEITY	COMPOSITION	SEPARATION
Element				
Compound				
Mixture				

2 Stainless steel and sterling silver are two commonly used alloys. Draw a diagram of each. Discuss the uses of each and describe the reason for alloying instead of using the pure metal.

3 In cold climates, salt is spread on icy roads and antifreeze is placed in car engines. Why is this so? How does this relate to distinguishing between fresh and salt water?

4 $2Pb(NO_3)_2(s) \xrightarrow{\text{Heat}} 2PbO(s) + 4NO_2(g) + O_2(g)$

a What observations would indicate that this chemical reaction had occurred?

b What safety precautions would need to be observed to carry out this chemical reaction?

c What are the compounds involved in this chemical equation?

d What is the element involved in this chemical reaction?

e What type of mixture is present in the reaction vessel?

5 $2C_8H_{18}(l) + 25O_2(g) \rightarrow 16CO_2(g) + 18H_2O(l)$

a What are the compounds involved in this chemical reaction?

b Which application would use this chemical reaction?

c What information is missing from this chemical equation?

d What are the short-term problems with this chemical reaction?

e What are the long-term problems with this chemical reaction?

f Use the chemical reaction in question **5** showing the combustion of petrol to write a balanced chemical reaction for the combustion of:

i natural gas

ii diesel.

6.3 Separation methods

1 Complete a summary in the following table to show the main separation methods used in school laboratories.

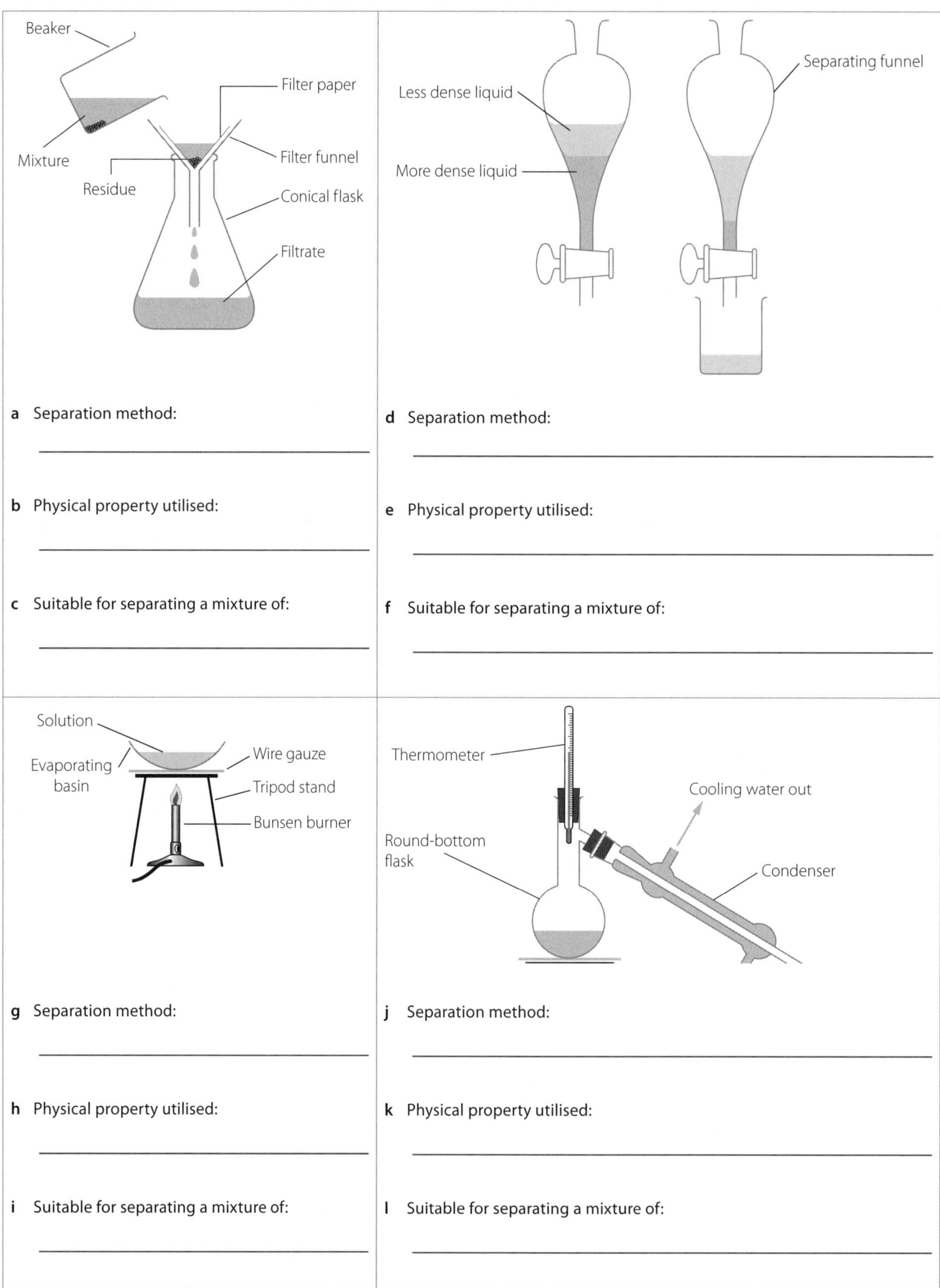

a Separation method: ______	**d** Separation method: ______
b Physical property utilised: ______	**e** Physical property utilised: ______
c Suitable for separating a mixture of: ______	**f** Suitable for separating a mixture of: ______
g Separation method: ______	**j** Separation method: ______
h Physical property utilised: ______	**k** Physical property utilised: ______
i Suitable for separating a mixture of: ______	**l** Suitable for separating a mixture of: ______

9780170412391

6.4 Real world distillation

The research article *African and Australian insect repellents for malaria prophylaxis* by Sarah Windsor and Amanda Neilen was published in 2014. This investigation determined the concentrations of volatile botanical active ingredients in insect repellents from Africa. Read this excerpt from the materials and methods section, and answer the following questions.

> Between 20 g and 30 g of African insect repellent was squeezed from its tube into a 1 L round-bottom flask. 600 mL of MilliQ water and a few boiling chips were added to the flask. 50 mL of hexane which contained 15 mg of n-hexadecane internal standard was pipetted into the flask and acted as a solvent trap for volatile botanic ingredients. Each sample was hydro-distilled on a Barnstead Electrothermal apparatus for 4 hr with the heating mantle set to level 5, then cooled to room temperature over a period of 2 hr. Two 1.5 mL aliquots were transferred by glass Pasteur pipette from the condenser to GC-MS vials for analysis.

1 Which solvent is denser?

____

____

2 Which solvent is less dense?

____

____

3 Which solvent stays in the round bottom flask?

____

____

4 Which solvent is transferred to the condenser?

____

____

5 Which solvent are the volatile botanical ingredients more soluble in?

____

____

6 Which solvent is the n-hexadecane internal standard more soluble in?

____

____

7 Draw a diagram of the equipment and consumables at the beginning of the separation.

8 Draw a diagram of the equipment and consumables at the end of the separation.

9 Why was a heating mantle used instead of a Bunsen burner, tripod and gauze mat in this experiment?

____

____

10 Why were two aliquots taken from the condenser?

____

____

6.5 Real world separatory funnel use

The research article *African and Australian insect repellents for malaria prophylaxis* by Sarah Windsor and Amanda Neilen was published in 2014. This investigation determined the concentration of *N,N*-diethyl-m-toluamide (DEET) active ingredient in insect repellents from Australia. Read this excerpt from the materials and methods section and answer the following questions.

> ... approximately 1 g of [Australian insect repellent] sample was scratched from the stick into a separatory funnel. ... 100 mL MilliQ water was poured into the separatory funnel. 50 mL of hexane which contained 100 mg of n-hexadecane internal standard was pipetted into the funnel. The funnel was inverted and shaken five times and left for 30 min. Approximately five drops were transferred by glass Pasteur pipette from the upper hexane layer to a 1.5 mL aliquot GC-MS vial and diluted with hexane prior to analysis. The lower water layer was emptied out of the separatory funnel tap into the collection beaker. The emulsion layer was broken up into hexane and water layers via vacuum filtration. The filtrate was poured into the separatory funnel and the lower water layer was emptied out of the separatory funnel tap into the collection beaker. ... The hexane layer was emptied out of the separatory funnel tap into a hexane waste container. The water collected was poured from the beaker into the separatory funnel. Eight extractions in total were performed.

Windsor SAM, Neilen A (2014) African and Australian Insect Repellents for Malaria Prophylaxis. JSM Chem 2(1): 1008. © 2014 Windsor et al.

1 Which solvent is denser?

____

____

2 Which solvent is less dense?

____

____

3 Which solvent is the DEET active ingredients more soluble in?

____

____

4 Which solvent is the n-hexadecane internal standard more soluble in?

____

____

5 Why was the collected sample diluted prior to analysis?

____

____

6 Draw a diagram of the three layers in the separatory funnel.

7 Draw a flow chart of this process.

8 Why do eight extractions need to be performed?

9 How much DEET is in the eighth collected sample?

10 How is the total concentration of DEET determined?

6.6 Real world liquid–liquid extractions

The research article *Chemical profiles of honeys originating from different floral sources and geographic locations examined by a combination of three extraction and analysis techniques* was written by Daniel Meloncelli, Sarah Windsor and P. Brooks and published in 2015. This investigation determined the concentration of active ingredients in Leatherwood and Manuka honeys from Australia and New Zealand. Read this excerpt from the materials and methods section and answer the following questions.

> Homogenised and subsampled honey (2 g) was weighed into a 16 mm x 100 mm test tube. Ethyl acetate [ethanoate] (2 mL), which contained n-hexadecane internal standard (107 µg/mL), was added to the honey. The test tube was covered with aluminium foil and placed into a 50 °C water bath for 5 min. The sample was vortexed … vigorously for 3 min and returned to the water bath for 5 min. This was repeated a further two times. The sample was settled for 15 min. The ethyl acetate supernatant was collected with a Pasteur pipette and transferred to a vial and dried with anhydrous Na_2SO_4 [sodium sulfate] (50 mg). One ethyl acetate supernatant (100 µL) was transferred to a 300 µL insert in a new 2 mL vial for GC-MS analysis. A second 100 µL of ethyl acetate supernatant was transferred to a 300 µL insert containing 50 µL of [bis(trimethylsilyl) trifluoroacetamide] BSTFA in a 2 mL vial and heated to 65 °C for 1 hr prior to GC-MS analysis.

Meloncelli, D., Windsor, S., & Brooks, P. (2015). CHEMICAL PROFILES OF HONEYS ORIGINATING FROM DIFFERENT FLORAL SOURCES AND GEOGRAPHIC LOCATIONS EXAMINED BY A COMBINATION OF THREE EXTRACTION AND ANALYSIS TECHNIQUES. Journal Of Fundamental And Applied Sciences, 7(2), 169-184. doi:http://dx.doi.org/10.4314/jfas.v7i2.3

1 Which solvent is denser?

____

____

2 Which solvent is less dense?

____

____

3 What is the effect of heating on solubility?

____

____

4 What is the effect of stirring on solubility?

____

____

5 Why were the heating and stirring repeated three times?

____

____

6 Why was BSTFA added to one of the ethyl acetate supernatants?

____

____

7 Why was the BSTFA/ethyl acetate supernatant heated?

____

____

8 Why was the BSTFA/ethyl acetate supernatant left for 1 hr?

____

____

9 Draw a diagram of the two layers in the test tube.

10 Draw a flow chart of this process.

1 Copy and complete the materials classification triangle below.

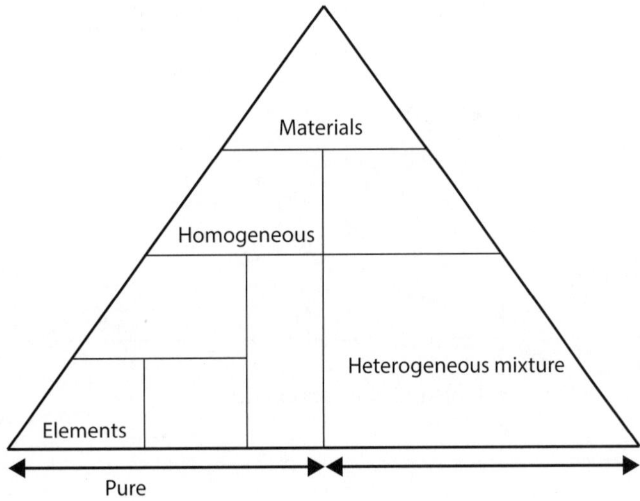

2 $Cu^{2+}(aq) + 2NaOH(aq) \rightarrow 2Na^{+}(aq) + Cu(OH)_2(s)$

a What observations would suggest the above reaction had occurred?

____

b What type of mixture is present in the reaction vessel?

____

c What compound is produced in this reaction?

____

d What type of separation technique could be used to isolate the compound produced in this reaction?

____

____

3 $CuO(s) + 2H^{+}(aq) \rightarrow Cu^{2+}(aq) + H_2O(l)$

a What observations would suggest the above reaction had occurred?

____

b What type of mixture is present at the end of the reaction?

____

c What compound is a reactant in this reaction?

____

d What compound is a product in this reaction?

4 The following questions relate to the structures of vitamin B2 and vitamin D2 below.

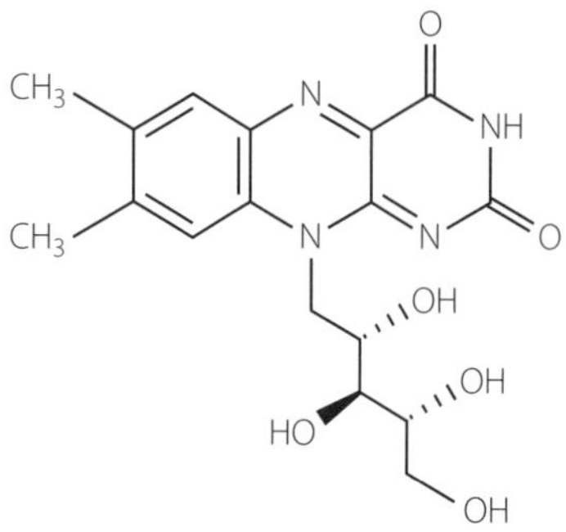

Vitamin B2

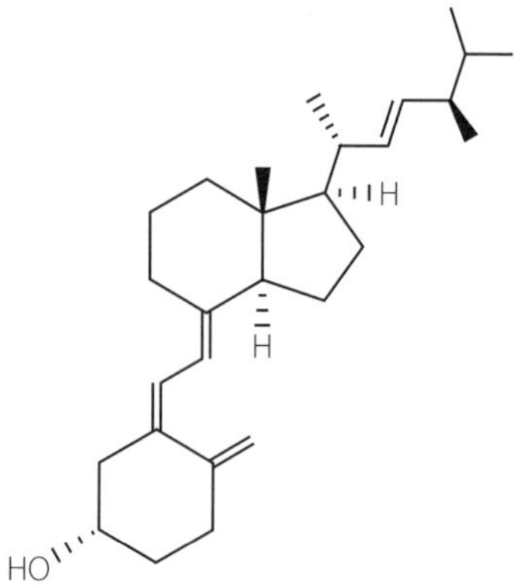

Vitamin D2

a Which vitamin is fat soluble?

i Discuss why this vitamin is fat soluble in terms of polarity.

ii Identify the parts of this vitamin's structure that impart its fat solubility.

iii Does this vitamin need to be consumed by humans on a regular basis and why?

b Which vitamin is water soluble?

i Discuss why this vitamin is water soluble in terms of hydrophilicity.

ii Identify the parts of this vitamin's structure that impart its water solubility.

iii Does this vitamin need to be consumed by humans on a regular basis and why?

9780170412391

c If a biological sample contained both vitamins, describe a process by which these vitamins could be separated in terms of technique, solvents and density of layers.

d If I was a representative from a pharmaceutical company, which vitamin would be easier to sell to the public and why?

5 High brass contains 65% copper and 35% zinc. Carbon steel contains 99% iron and 1% carbon. Stainless steel contains 88% iron, 11% chromium and 1% carbon. The table below displays the atomic radii of elements involved in these alloys.

ELEMENT	ATOMIC RADIUS (Å)
Carbon	0.7
Chromium	1.40
Copper	1.35
Iron	1.40
Zinc	1.35

a Which alloy is a substitution alloy and why?

b Which alloy is an interstitial alloy and why?

c Which alloy is a substitution and interstitial alloy and why?

7 Bonding and properties

Summary

- Metallic substances consist of positive ions surrounded by delocalised electrons. The structure of metals results in their characteristic properties:
  - conductivity of heat and electricity when solid
  - malleability
  - ductility.
- Ionic compounds consist of positive and negative ions held together by electrostatic interactions in a crystalline lattice. The structure of ionic compounds results in their characteristic properties, such as:
  - high melting point
  - brittleness
  - ability to conduct electricity when molten or in solution.
- Covalent molecular compounds consist of covalently bonded molecules, which are held together by comparatively very weak intermolecular forces. The structure of covalent molecular compounds results in their characteristic properties such as:
  - not conducting electricity
  - formation of soft, brittle solids
  - low melting and boiling points.
- Hydrocarbons are covalent molecular compounds made from hydrogen and carbon atoms. Hydrocarbons include alkanes, alkenes and aromatic compounds.
  - Alkanes are saturated hydrocarbons, with single covalent bonds only
  - Alkenes are unsaturated hydrocarbons with one double covalent bond between carbon atoms per molecule. The C $\leftrightarrow$ C double bond in an alkene results in alkenes taking part in addition reactions, which alkanes cannot do.
  - Aromatic compounds contain benzene rings, hexagonal rings of six carbon atoms covalently bonded to each other and to a hydrogen atom. The level of unsaturation is not as great as for an alkene.
- Alkanes, alkenes, aromatics and all other organic compounds can be named systematically, so that the structure can be deduced from the name and vice versa.
- Covalent network compounds consist of covalently bonded atoms but in a network arrangement with no discrete molecules. Common examples include diamond and silicon dioxide. The structure of covalent network compounds results in properties including:
  - very high melting and boiling points
  - very hard substances
  - zero electrical or thermal conductivity.

- Covalent layer compounds consist of covalently bonded atoms but in a network arrangement with no discrete molecules. Common examples include diamond and silicon dioxide. The structure of covalent layer compounds results in properties including:
  - very high melting and boiling points
  - very hard substances
  - zero electrical or thermal conductivity.

7.1 Important terms

1 Complete the following crossword using the clues on page 66.

1	2	3	4	5	6	7	8	9	10	11	12	13	14	15	16	17	18	19			

9780170412391

Across	**Down**
8 A substance that can be repeatedly deformed, or hammered, without shattering **11** Can be drawn out to form a thin wire **12** An organic compound where each carbon is bonded to four different atoms, without any double bonds being present **13** Compounds containing a benzene ring **14** __________ strength. The ability of a material to resist a stretching force **15** Mass per unit volume **16** A reaction where two molecules combine, resulting in a larger molecule being formed. **17** _____________ isomers – two compounds containing the same number of each atom, but with these atoms arranged in a different way	**1** A carbon atom (or group of carbon atoms) that is not part of the longest chain of carbon atoms in an organic molecule **2** Electrons that exist within a structure but are not bound to an individual atom **3** Bonding within molecules **4** A reaction where one atom of a compound is replaced by an atom of a different element **5** ___________ chemistry – the study of carbon-based compounds **6** Resistance to being scratched **7** Organic compounds containing at least one double bond between carbon atoms **9** Compounds made from hydrogen and carbon only **10** A substance which will shatter rather than bend, when a force is applied **18** A reaction where a substance is burnt in oxygen **19** Different forms of the same element, with different physical properties

2 Write a short summary about bonding and the properties of substances using the terms in the crossword.

____

____

____

____

____

7.2 Bonding within solids

1 Complete the following table to compare and contrast the bonding and properties of different types of solids.

SOLID TYPE	EXAMPLE	DESCRIPTION OF STRUCTURE
Ionic		
Covalent molecular		
Covalent network		
Covalent layer		
Metallic		

7.3 Drawing and naming hydrocarbons

1 Draw the following structures in the space provided.

a Pentane

b 2,3-Pentadiene

c Benzene

9780170412391

d 3-Hexene

e 2-Methylbutane

f 2,2,4-Trimethylpentane

2 Name and draw five isomers with the molecular formula C_5H_{10}, all of which are alkenes.

a Isomer name: ______________________________

b Isomer name: ______________________________

c Isomer name: ______________________________

9780170412391

d Isomer name: ______________________________

e Isomer name: ______________________________

7.4 Reactions of hydrocarbons

1 Write balanced equations for the following reactions.

a Combustion of hexane

____

____

b 2-Methyl-1-pentene with hydrogen gas

____

____

c 2-Butene with water

____

____

d Benzene with bromine

____

____

e 2-Pentene with bromine

____

____

EVALUATION

1 A substance has the following properties:

- conducts electricity when molten
- solid at room temperature
- melts on heating with a Bunsen burner.

Circle the best option to show which bonding types are likely to be present in the substance.

A Metallic or ionic

B Ionic or covalent

C Metallic or covalent

D Metallic or ionic or covalent

2 Complete the table by using the evidence given to indicate whether the structure of each substance is covalent molecular, ionic or metallic.

DESCRIPTION	MELTING POINT	CONDUCTIVITY WHEN SOLID	CONDUCTIVITY WHEN MOLTEN	COVALENT MOLECULAR, IONIC OR METALLIC?
Silvery solid	98°C	Yes	Yes	
White solid	920°C	No	Yes	
Shiny solid	Sublimes (vaporises immediately on forming a liquid)	No	–	
Yellow solid	110°C	No	No	

3 Provide examples to explain the difference between the following terms:

a Allotrope and isotope

______

______

______

b Covalent molecular and covalent network structure

______

______

______

4 Using International Union of Pure and Applied Chemistry (IUPAC) conventions, state the correct name for the following organic chemicals.

a Hexan-6-ol

____

b 4,4-Dibromopentane

____

c 1-Propylmethane

____

8 Chemical reactions

LEARNING

Summary

- Chemical reactions involve the creation of new substances. During chemical reactions, bonds are broken and (or) new bonds are formed.
- Chemical reactions involve energy changes. These energy changes lead to one or more of the following observable events:
  - colour change
  - change in smell
  - production of new gases or vapour
  - formation of a precipitate
  - temperature change
  - absorption or emission of radiation (light)
  - evolution of sound
  - difficulty in reversing the change.
- Phase changes are changes of state, meaning a transformation between the solid, liquid or gaseous state. Melting, freezing, vaporisation (evaporation and boiling) condensation, sublimation and deposition are all changes of phase and are normally associated with a change of temperature.
- Chemical reactions can be represented by chemical equations. In a chemical equation, the reactants (the substances that exist before a chemical change (or reaction) takes place) are written on the left-hand side of the arrow. The products (the new substances formed) are written on the right-hand side of the arrow. The single arrow always points in the direction of the reaction; that is, from the reactants to the products.
- Chemical equations can be either balanced complete chemical equations or balanced net ionic chemical equations. Numbers (coefficients) are written in front of chemical formulas to balance the equation so that there is the same number of atoms of each element on each side.
- Spontaneous reactions are those that occur without any addition of energy from the surroundings; many spontaneous reactions occur at room temperature and below. Spontaneous reactions can occur at any speed; some are fast, some very slow.
- Some metals react spontaneously with water, acid and oxygen. Corrosion is the spontaneous reaction of a metal with oxygen in a moist environment.
- The more reactive metals are more likely to react spontaneously.

8.1 Chemical reactions

A physical change is a change in size, shape or state; no new substance is formed. A chemical change is a change in the physical and chemical properties; a new substance is formed. Identify the following as either a physical or a chemical change. (Circle the correct classification.)

1 An ice cube is placed in the sun, a puddle of water forms; later the puddle disappears. *Physical/chemical*

2 Two chemicals are mixed together and a gas is produced. *Physical/chemical*

3 A steel rake lying in the garden changes colour as it rusts. *Physical/chemical*

4 Peppercorns are crushed to a powder. *Physical/chemical*

5 Magnesium ribbon is placed in a Bunsen flame and a bright white light is produced. *Physical/chemical*

6 A piece of sodium is dropped into a beaker of water and a bright yellow flame is observed. *Physical/chemical*

7 Salt (NaCl) is added to water and the solid disappears. *Physical/chemical*

8 Chocolate powder is dissolved in milk. *Physical/chemical*

9 A marshmallow is toasted over a campfire. *Physical/chemical*

10 An apple is sliced in half. *Physical/chemical*

11 An electrical spark ignites gas on a gas stove to produce a flame. *Physical/chemical*

12 A seed germinates and grows into a seedling. *Physical/chemical*

8.2 Balancing chemical equations

WORKED EXAMPLE

Chemical reactions can be represented by a:

1 word equation.

2 chemical equation, including physical states (s, l, g, or aq).

3 balanced chemical equation.

For example, the reaction of carbon burning in air (combustion) to form carbon monoxide can be represented as follows. (Hint: if the coefficient is 1, it does not need to be included.)

1 Carbon + oxygen → carbon monoxide

2 $C(s) + O_2(g) \rightarrow CO(g)$

3 $2C(s) + O_2(g) \rightarrow CO(g)$

1 For each of the following descriptions, represent the reaction in the three ways described in the Worked Example on page 74.

a Solid aluminium oxide is formed when aluminium metal is burned in air. (Hint: air contains oxygen – remember diatomic elements.)

b Water is formed from the combustion of hydrogen gas in air.

c When a piece of zinc metal is placed in a solution of hydrochloric acid (HCl), hydrogen gas and a solution of zinc chloride solution is formed.

d Aqueous potassium hydroxide (KOH) reacts with aqueous sulfuric acid (H_2SO_4) to produce a solution of potassium sulfate together with the formation of water.

e A solution of magnesium chloride reacts with a solution of sodium hydroxide to produce magnesium hydroxide as a solid precipitate in a solution of sodium chloride.

f Barium sulfate precipitate is formed in a solution of potassium nitrate when a solution of barium nitrate is added to a solution of potassium sulfate.

g Nitrogen heated with hydrogen produces ammonia. (Hint: remember diatomic elements.)

____

____

____

2 Balance the following chemical equations as shown in the following example.

Unbalanced:

$\_C_2H_4 + \_O_2 \rightarrow \_CO_2 + \_H_2O$

Balanced:

$\_C_2H_4 + \underline{3}O_2 \rightarrow \underline{2}CO_2 + \underline{2}H_2O$

a $\_Li + \_I_2 \rightarrow \_LiI$

b $\_NaClO_3 \rightarrow \_NaCl + \_O_2$

c $\_K_3PO_4 + \_HCl \rightarrow \_KCl + \_H_3PO_4$

d $\_C_5H_{12} + \_O_2 \rightarrow \_CO_2 + \_H_2O$

e $\_Ca(NO_3)_2 + \_Na_3PO_4 \rightarrow \_Ca_3(PO_4)_2 + \_NaNO_3$

f $\_C_6H_8 + \_O_2 \rightarrow \_CO_2 + \_H_2O$

g $\_H_3PO_4 + \_KOH \rightarrow \_K_3PO_4 + \_H_2O$

h $\_Al(OH)_3 + \_H_2CO_3 \rightarrow Al_2(CO_3)_3 + \_H_2O$

i $\_FeS_2 + \_O_2 \rightarrow \_Fe_2O_3 + \_SO_2$

j $\_Ca_3(PO_4)_2 + \_SiO_2 + \_C \rightarrow \_CaSiO_3 + \_CO + \_P$

8.3 Chemical reactivity

Metals have different reactivities – metals that are more reactive can displace metals that are less reactive from their compounds formed by combining with other elements. The more reactive metals (as shown in Table 8.3.1) will be more likely to undergo spontaneous reactions.

TABLE 8.3.1 Reactivity of some metals and halogens

METAL ACTIVITY SERIES	HALOGEN REACTIVITY
Li – **Most reactive**	F_2 – **Most reactive**
Rb	Cl_2
K	Br_2
Ba	I_2 – **Least reactive**
Ca	
Na	
Mg	
Al	
Mn	
Zn	
Cr	
Fe	
Ni	
Sn	
Pb	
H_2	
Cu	
Hg	
Ag	
Pt	
Au – **Least reactive**	

The following metals are listed in order of reactivity (most reactive first):

sodium > magnesium > zinc > copper

1 Complete the following table for each of the given metals.

	SODIUM	MAGNESIUM	ZINC	COPPER
Dropped into water				
Heated				
Added to dilute hydrochloric acid				

9780170412391

2 Which of the four metals would be suitable for making saucepans? Explain why the others are not.

3 Below is a list of metals in order of decreasing reactivity. **Q** and **R** are unidentified metals.

K > **Q** > Ca > Mg > Al > Zn > **R** > Fe > Cu

a Will **Q** react with cold water?

b Will **R** react with cold water?

c Will **R** react with dilute hydrochloric acid?

d Will **R** displace copper from copper sulfate in solution?

4 Decide whether the statements following are True (T) or False (F). If false, provide the correct statement.

a Spontaneous chemical reactions require the input of additional energy such as heat, light or electricity. T F

b During chemical reactions, bonds are broken and new bonds are formed. T F

c All spontaneous reactions occur at the same rate. T F

d Metals can react spontaneously. T F

e Corrosion is the spontaneous reaction of a metal with oxygen in a moist environment. T F

____

____

f When ionic compounds are dissolved in water, they can be broken up (dissociated) into their constituent ions. T F

____

____

5 Some metals react spontaneously with water, acid and oxygen. The generalised equations for these reactions are given below. Provide one example for each class of reaction and represent it as a balanced chemical equation, as shown in the following example.

Reaction class: Metal + acid $\rightarrow$ a salt + hydrogen

Example: $Cd + H_2SO_4 \rightarrow CdSO_4 + H_2$

a Metal + water $\rightarrow$ a salt + hydrogen

Example:

____

b Metal + acid $\rightarrow$ a salt + hydrogen

Example:

____

c Metal + oxygen $\rightarrow$ metal oxide

Example:

____

6 Write net ionic equations for each of the following reactions and explain why they are expected to occur spontaneously.

a $Zn(s) + CuSO_4(aq) \rightarrow Cu(s) + ZnSO_4(aq)$

____

____

____

____

____

b $K_2CO_3(aq) + CaCl_2(aq) \rightarrow CaCO_3(s) + 2KCl(aq)$

7 Write net ionic equations to represent the following reactions.

a Magnesium metal and water

b Zinc and hydrochloric acid

c Iron and sulfuric acid

d Nickel nitrate solution and sodium iodide solution to form sodium nitrate solution and solid lead iodide.

EVALUATION

1 Classify and describe each of the following changes as either chemical or physical.

a Wire is bent to form a circular shape.

____

b Wood is burnt in the fireplace.

____

c A liquid is heated until it boils and evaporates.

____

d Zinc metal pieces are placed in sulfuric acid, the metal dissolves and bubbles of a colourless gas are produced.

____

e Sugar is dissolved in warm water to form syrup.

____

f A lighted taper is applied to a balloon filled with hydrogen and it explodes.

____

2 Balance the following chemical equations.

a $\_\_N_2 + \_\_O_2 \rightarrow \_\_N_2O$

b $\_\_NaI + \_\_Cl_2 \rightarrow \_\_NaCl + \_\_I_2$

c $\_\_C_3H_8 + \_\_O_2 \rightarrow \_\_CO_2 + \_\_H_2O$

d $\_\_Mg(NO_3)_2 + \_\_K_3PO_4 \rightarrow \_\_Mg_3(PO_4)_2 + \_\_KNO_3$

3 Which of the following is the least active metal? (Circle the correct answer.)

A Aluminium

B Zinc

C Gold

D Copper

E Lead

4 Write balanced chemical equations for the following reactions.

a Combustion of hexane (C_6H_{14}) to yield carbon dioxide and water.

____

____

b Rusting of iron when it reacts with oxygen and water to form iron(II) hydroxide.

____

____

c Reaction of nitrogen with hydrogen to produce ammonia.

____

____

9 Exothermic and endothermic reactions

Summary

- In a chemical reaction bonds are broken in the reactants, and bonds are formed to make the products. These changes involve energy.
- Energy is required to break bonds to release the atoms (or groups of atoms) in the molecule so that they can undergo a subsequent reaction. Therefore, energy needs to be put in to break the bonds in the reactants.
- The law of conservation of energy states that energy cannot be created or destroyed; it can only be changed from one form to another.
- If more energy is stored in the bonds of the reactants than the products, then when the products are formed, the extra energy is released as heat, sound or light.
- In an exothermic reaction, energy is released because the energy of the reactants is greater than the energy of the products. In an endothermic reaction, energy is absorbed because the energy of the reactants is less than the energy of the products.
- Physical processes, such as changes of state, can also involve the release or input of energy.
- Enthalpy is the heat absorbed in a chemical reaction at constant pressure. The change in enthalpy, also referred to as the heat of reaction, is given the symbol ΔH. It is measured in kilojoules (kJ).
- Endothermic reactions involve the input (absorption) of energy, hence they are characterised by positive ΔH values. Exothermic reactions involve the release of energy; hence they are characterised by negative ΔH values.
- Temperature is a measure of the average total kinetic energy of the atoms and molecules in a system.
- In molecular solids, liquids and gases, kinetic energy is comprised of translational, rotational and vibrational motion. Atoms do not contain rotational kinetic energy but atoms in the liquid or solid state can contain a form of vibrational energy which is due to the attractive interatomic forces holding the liquid or solid together in that state.
- The chemical bond energy (or bond enthalpy) is the energy associated with the attractive and repulsive electrostatic interactions between electrons and nuclei. A specific bond energy (or enthalpy) is defined by the amount of energy required to break one mole of the stated bond to give separated atoms. The units are $kJ\,mol^{-1}$.
- Temperature (°C) = Temperature (K) −273
- The specific heat capacity (C) is the amount of heat q (kJ) needed to increase the temperature of 1 g of a substance by 1 K.

9780170412391

- The amount of heat gained or lost in a chemical reaction (q) is directly proportional to the:
  - specific heat capacity (C) of the substance
  - mass in moles (m) of the substances reacting
  - change in temperature (ΔT).

  Expressed mathematically, q is dependent on the product of these three factors:

$$q = mC\Delta T \text{ (units kJ)}$$

- The enthalpy change associated with the chemical reaction is the heat gained or lost per mole of reactant; hence, it is obtained by dividing q by the number of moles of reactant:

$$\Delta H = \frac{q}{m} \text{ (units kJ mol}^{-1}\text{)}$$

- Hess's Law states that the enthalpy change accompanying a chemical change is independent of the route by which the chemical change occurs. Regardless of any intermediates in the reaction:

$$\Delta H = H_{\text{products}} - H_{\text{reactants}}$$

- Combustion reactions are reactions where substances are burnt in the presence of oxygen and release large amounts of energy. The enthalpy change ΔH involved in combustion reactions is usually called the heat of combustion.

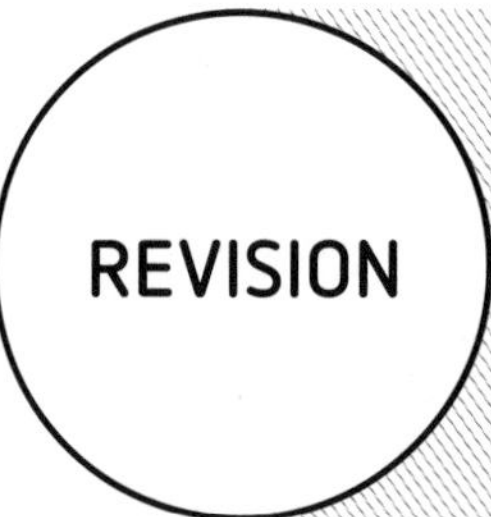

9.1 Important terms

1 Complete the crossword using the clues on page 86.

 9780170412391

Across	Down
4 Sign for ΔH in an exothermic reaction	**1** Sign for ΔH in an endothermic reaction
5 The term used to define the average total kinetic energy of the atoms and molecules in a system	**2** Required to break bonds in the reactants for a chemical reaction to occur
6 Type of reaction where energy is gained because the energy of the products is greater than the energy of the reactants	**3** Type of reactions where substances are burnt in the presence of oxygen and release large amounts of energy
10 Name of scientist after whom the law is named whereby the enthalpy change accompanying a chemical change is independent of the route by which the chemical change occurs	**5** A type of motion contributing substantially to the kinetic energy of atoms or molecules in gases but absent in solids
11 The type of motion contributing to the major source of kinetic energy in solids	**7** Type of reaction where energy is released because the energy of the reactants is greater than the energy of the products
13 A type of motion contributing to the kinetic energy of atoms or molecules in gases and liquids but essentially absent in solids and undefined for atoms because they are spherically symmetrical	**8** Name of temperature scale used to measure temperature relative to absolute zero
	9 The term given for the heat absorbed in a chemical reaction
	12 Name of temperature scale used to measure temperature relative to the freezing point of water, being assigned as zero

2 Write a short summary of exothermic and endothermic reactions using the terms in the crossword.

9.2 Exothermic and endothermic reactions

1 Classify the following processes as exothermic (Exo) or endothermic (Endo).

PROCESS	EXO OR ENDO	PROCESS	EXO OR ENDO
Water vapour condensing to liquid water (e.g. clouds turning to rain)		Conversion of frost to water vapour	
Producing sugar and O_2 by photosynthesis		Burning gas or petrol (combustion)	
Mixing sodium hydroxide pellets with water		Mixing water with ammonium nitrate	
Mixing water with strong acids		Melting of ice cubes or snow	
Splitting apart a gas molecule (e.g. O_2 or N_2)		Mixing water with cooking salt	
Snow formation in clouds		Baking bread	
Burning sugar		Cooking an egg	
Nuclear fission		Making ice cubes in a freezer	
The evaporation of water		Rusting of iron in air	

2 Reactions involve either or both bond breaking and bond making. This equation below using molecular models shows what happens when methane (CH_4) burns in oxygen (O_2).

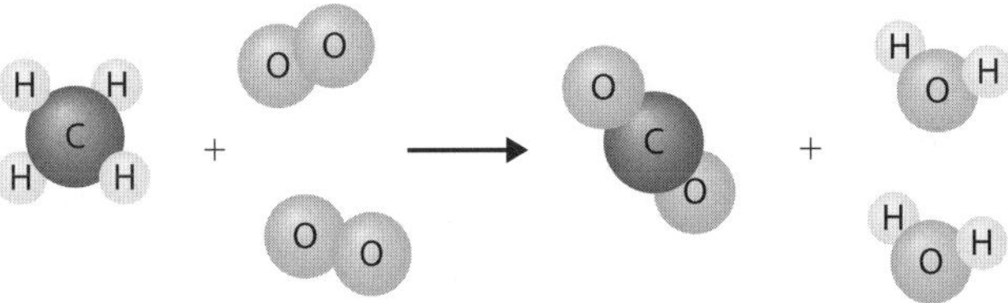

a On the figure, mark the bonds broken in red and the bonds formed in green.

b Write the equation using chemical symbols.

____

____

c Complete the table to show the number of bonds broken and formed.

BONDS BROKEN	NUMBER	BONDS FORMED	NUMBER
Between carbon and hydrogen		Between carbon and oxygen	
Between oxygen atoms		Between hydrogen and oxygen	

d Conclude whether the reaction is exothermic or endothermic overall.

____

____

9780170412391

3 Write balanced thermochemical equations for the following reactions.

For example, when carbon dioxide and water are formed from benzene (liquid, C_6H_6) and oxygen gas, 3267 kJ of heat is released:

$$2C_6H_6(l) + 15O_2(g) \rightarrow 12CO_2(g) + 6H_2O(l)\ \Delta H = -3267\,kJ$$

a Carbon dioxide and water are formed from propane gas (C_3H_8) and oxygen gas in a reaction that releases 2220 kJ heat.

____

____

b Ammonia gas (NH_3) absorbs 92 kJ of heat to form nitrogen and hydrogen gases.

____

____

c Calcium oxide solid is formed when 180 kJ of heat is absorbed by solid calcium carbonate. Carbon dioxide is the other product.

____

____

d Ammonia gas reacts with oxygen to form water vapour and nitrogen oxide (NO) gas. The reaction releases 905 kJ of energy.

____

____

e Liquid water freezes to form ice. The latent heat of fusion is 6.02 kJ mol^{-1}.

____

____

4 When all reactants and products are in their standard states (denoted °), the enthalpy change accompanying a reaction ($\Delta H^{o}_{reaction}$) is the sum (symbolised by Σ) of the standard heats of formation for all of the products ($\Sigma\Delta H^{o}_{f(products)}$) minus the sum of the standard heats of formation for all of the reactants ($\Sigma\Delta H^{o}_{f(products)}$):

$$\Delta H^{o}_{reaction} = \Sigma\Delta H^{o}_{f(products)} - \Sigma\Delta H^{o}_{f(reactants)}$$

For example:

$$NaOH(s) + HCl(g) \rightarrow NaCl(s) + H_2O(g)$$

Reactants		**Products**	
ΔH^{o}_{f} (kJ mol^{-1})		ΔH^{o}_{f} (kJ mol^{-1})	
NaOH	–426.7	NaCl	–411.0
HCl	–92.3	H_2O	–241.8
Total:	–519.0	Total	–652.8

Therefore, the change in enthalpy = –133.8 kJ mol^{-1}.

TABLE 9.2.1 Standard enthalpies of formation of some common compounds

COMPOUND	ΔH°_f (kJ mol^{-1})	COMPOUND	ΔH°_f (kJ mol^{-1})
$CH_4(g)$	−74.8	$HCl(g)$	−92.3
$CO(g)$	−110.53	$H_2O(g)$	−241.8
$CO_2(g)$	−393.5	$SO_2(g)$	−296.1
$NaCl(s)$	−411.0	$NH_4Cl(s)$	−315.4
$H_2O(l)$	−285.8	$NO(g)$	+90.4
$H_2S(g)$	−20.1	$NO_2(g)$	+33.9
$H_2SO_4(l)$	−811.3	$SnCl_4(l)$	−545.2
$MgSO_4(s)$	−1278.2	$SnO(s)$	−286.2
$MnO(s)$	−384.9	$SnO_2(s)$	−580.7
$MnO_2(s)$	−519.7	$SO_2(g)$	−296.1
$NaCl(s)$	−411.0	$SO_3(g)$	−395.2
$NaF(s)$	−569.0	$ZnO(s)$	−348.0
$NaOH(s)$	−426.7	$ZnS(s)$	−202.9
$NH_3(g)$	−46.2		

Use the standard enthalpies of *formation* table (see Table 9.2.1) to determine the change in enthalpy for each of the following reactions.

a $2CO(g) + O_2(g) \rightarrow 2CO_2(g)$

b $CH_4(g) + 2O_2(g) \rightarrow CO_2(g) + 2H_2O(l)$

c $2H_2S(g) + 3O_2(g) \rightarrow 2H_2O(l) + 2SO_2(g)$

d $2NO(g) + O_2(g) \rightarrow 2NO_2(g)$

5 Hess's law (Note that published thermochemical data can vary, depending on the source.)

a Given the following data:

$$S(s) + O_2(g) \rightarrow SO_2(g) \qquad \Delta H = -297\,kJ$$
$$2SO_3(g) \rightarrow 2SO_2(g) + O_2(g) \qquad \Delta H = +198\,kJ$$

calculate the enthalpy change (ΔH) for the following reaction:

$$2S(s) + 3O_2(g) \rightarrow 2SO_3(g)$$

b Given the following data:

$$Sn + Cl_2 \rightarrow SnCl_2 \qquad \Delta H = -325\,kJ$$
$$SnCl_2 + Cl_2 \rightarrow SnCl_4 \qquad \Delta H = -186\,kJ$$

calculate ΔH for the following reaction:

$$Sn + 2Cl_2 \rightarrow SnCl_4$$

c Given the following data:

$$O_2 \rightarrow 2O \qquad \Delta H = +495\,kJ$$
$$2O_3 \rightarrow 3O_2 \qquad \Delta H = -427\,kJ$$
$$NO + O_3 \rightarrow NO_2 + O_2 \qquad \Delta H = -199\,kJ$$

calculate ΔH for the following reaction:

$$NO + O \rightarrow NO_2$$

d Given the following data:

$$4Al + 3O_2 \rightarrow 2Al_2O_3 \qquad \Delta H = -3352\,kJ$$

$$Mn + O_2 \rightarrow MnO_2 \qquad \Delta H = -521\,kJ$$

calculate ΔH for the following reaction:

$$4Al + 3MnO_2 \rightarrow 2Al_2O_3 + 3Mn$$

____

____

____

6 The energy released when fuels are burned can be estimated using temperature and mass changes measured in calorimetric experiments.

a For each of the fuels methanol, ethanol, propanol and butanol, 1 g was placed in a glass burner under a calorimeter containing 200 cm^3 of water. The fuels were each burnt to completion and the temperature rise of the calorimeter was recorded. The results are shown in the following table.

Calculate each temperature rise and state which fuel releases most energy per gram.

FUEL	TEMPERATURE AT START (°C)	TEMPERATURE AT END (°C)	TEMPERATURE RISE (°C)
Methanol	23	54	
Ethanol	19	54	
Propanol	26	63	
Butanol	21	61	

____

____

____

____

____

____

____

____

____

____

b Each of the fuels methanol, ethanol, propanol and butanol were again placed in a glass burner under the calorimeter (containing 200 cm^3 of water). This time the fuels were weighed before and after each test. The tests were carried out by letting each fuel burn until the calorimeter water temperature had risen by precisely 10.0°C. The results are shown in the following table. Calculate the mass of each fuel required to release the same amount of energy, and state which fuel must release more energy per gram.

FUEL	MASS OF BURNER AT START (g)	MASS OF BURNER AT END (g)	MASS OF FUEL USED (g)
Methanol	145.5	141.9	
Ethanol	202.1	199.3	
Propanol	177.5	175.1	
Butanol	226.3	224.1	

7 A sample of a silvery grey metal was provided to an analytical chemist who was told that it was either bismuth (specific heat 0.122 $J g^{-1} °C^{-1}$) or cadmium (specific heat 0.232 $J g^{-1} °C^{-1}$). They measured out a 250 g sample of the metal, heated it to 96.0°C and then added it to 98.5 g of water at 21.0°C in a perfect calorimeter. The final temperature in the calorimeter was 30.3°C. Use the data to calculate the specific heat of the metal sample and then identify the metal.

The enthalpy changes that can occur during a chemical reaction are shown in the following graphs.

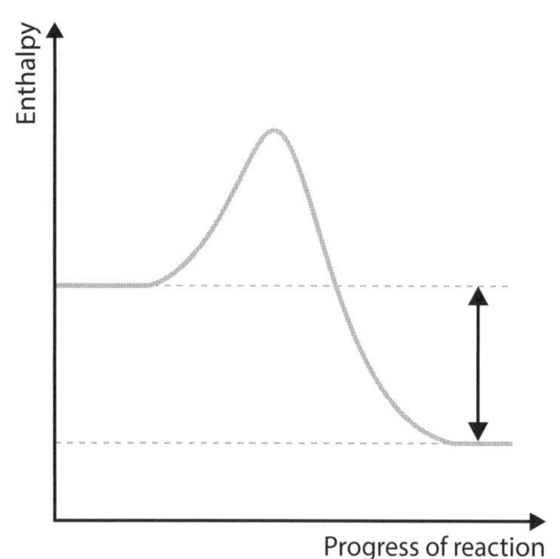

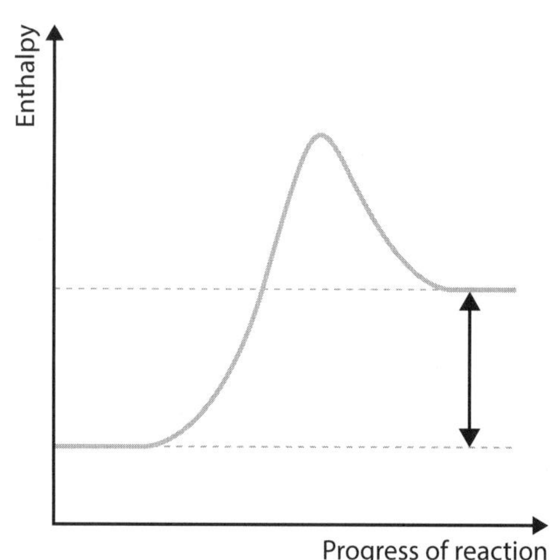

1 For each graph shown, complete the following tasks.

a Identify and label the 'Energy of reactants' and 'Energy of products'.

b On the y-axis, label the energy of reactants and products as 'ER' and 'EP' respectively.

c Identify and label the change in enthalpy (ΔH).

d Indicate whether the enthalpy change is positive or negative.

e Decide whether the graph is showing an exothermic or endothermic reaction and give each graph a title.

f Explain what the 'hump' on each graph is related to in terms of the reaction path.

____

____

____

g Explain why the exothermic reaction shown in the figure would not occur spontaneously.

____

____

____

9780170412391

h Describe how the graph for a spontaneous exothermic reaction would look different from the graphs for exothermic and endothermic reactions shown and sketch the graph.

2 The combination of coke and steam produces a mixture called 'coal gas', which can be used as a fuel or as a starting material for other reactions. Coke is essentially pure carbon; therefore, the equation for the production of coal gas is represented as follows.

$$2C(s) + 2H_2O(g) \rightarrow CH_4(g) + CO_2(g)$$

Determine the standard enthalpy change for this reaction from the following standard enthalpies of reaction:

$$C(s) + H_2O(g) \rightarrow CO(g) + H_2(g) \qquad \Delta H^{\circ} = +131.3\,kJ$$

$$CO(g) + H_2O(g) \rightarrow CO_2(g) + H_2(g) \qquad \Delta H^{\circ} = -41.2\,kJ$$

$$CH_4(g) + H_2O(g) \rightarrow 3H_2(g) + CO(g) \qquad \Delta H^{\circ} = +206.1\,kJ$$

3 A 30 g mass of a metal was heated to 97.0°C and then dropped into 100 mL of water at 24.0°C. The water sample was contained in an insulated chamber. The temperature of the water increased and reached an equilibrium at 26.0°C.

a How much heat did the water absorb?

b What quantity of heat was lost by the metal sample?

c Using the appropriate relationship for a calorimetric experiment, determine the heat capacity (JK^{-1}) of the 30 g metal sample.

d Hence, determine the specific heat capacity of the metal in part **c** ($Jg^{-1}K^{-1}$).

e The heat capacities of some common metals are given ($Jg^{-1}K^{-1}$):

Al, 0.92; Ni, 0.50; Cu, 0.38; Ag, 0.24; and Pb, 0.13

Identify the metal involved in the calorimetry experiment described in this question.

9780170412391

10 Measurement uncertainty and error

LEARNING

Summary

- Measurements, observations and recording of results are central to scientific investigation.
- In typical experiments, one independent variable is varied. A domain (i.e. a range of values) is chosen for the independent variable. Anything else that might influence any of the dependent variables (those that depend on the independent variable) is kept constant.
- Any experimental scientific report will present data based on measurements, observations and analyses of the data. The credibility of the conclusions presented in the report depends on the reliability of the data.
- A measurement of some quantity is carried out by taking a measuring standard and then using a measurement procedure to determine the measurement.
- The accuracy of a standard of measure needs to be determined before reporting a measurement using that standard.
- Any measuring device has an accuracy that can be determined only by comparison with a precisely known primary standard of measure.
- There is uncertainty in every measurement. A measurement is meaningful only when it is accompanied by a quote of the level of uncertainty (confidence factor) for the measurement.
- The limitations of a standard of measure, or limit of reading, are different for analogue and digital devices.
- The resolution (or limit of reading) is the minimum uncertainty in any measurement.
- Good precision means that all of the measurements lie within a small range. However, measurements may be a long way from the truth if the measurement, although precise, is inaccurate because of calibration error.
- Whenever possible, measurements should be repeated. For repeated measurements, the best estimate of the measured quantity is the average value (i.e. the mean, denoted $\bar{x}$).
- Repeated measurements will be randomly spread about the 'true value', and centred on that value.
- For fewer than 10 measurements, the best estimate of uncertainty is half the range.
- For a large number n of measurements, typically $n > 10$, the uncertainty is given by the standard deviation (σ) and the mean of the measurements is ($\bar{x}$):

$$\sigma = \text{standard deviation} = \sqrt{\frac{\sum_{i=1}^{n}(x_i - \bar{x})^2}{n-1}}$$

- Qualitative measurements or data are descriptions such as colour, formation of a precipitate or presence of bubbles.
- Quantitative data provide numerical measures of quantities which can be manipulated using mathematical operations.
- Always define the source of the uncertainty (e.g. reaction time, parallax, estimation between minimum markings and limit of reading) and quantify it. Do not use the term 'human error'!
- Always quote results to the correct number of significant figures. The number of significant figures displayed when quoting a measurement comes from estimating the uncertainty and estimating the consequences of any error propagation through calculations.
- The percentage uncertainty is a percentage calculated relative to the measured quantity.
- If there are more than a few data points, display them in a table.
- Graphs provide a powerful technique for identifying trends and relationships.

Guidelines

1 A measurement without units is meaningless.

2 A measurement quoted without an estimated range of uncertainty (or error) is also meaningless.

3 Always quote the final result of a measurement in the form:

$$X \pm \delta x$$

where X is the measured quantity and '$\pm \delta x$' is the uncertainty quoted to the same number of decimal places. For example, a measurement of the mass of a sample might given as:

$$\text{Mass} = (5.32 \pm 0.04)\,\text{g}$$

This expression means that the true mass is likely to lie between 5.28 g and 5.26 g.

If this quoted result is based on many repeated measurements, then it is implied that the probability of the mass being between 5.28 g and 5.36 g is 68% or within $\pm\sigma$ where $\pm\sigma = \pm 34\%$ is the standard deviation for a normal distribution.

4 Never quote errors to more than two significant figures because errors (i.e. uncertainties) are usually known only approximately. One significant figure is usually sufficient (e.g. 5.32 ± 0.04; or possibly 5.325 ± 0.035; but certainly not 5.325476 ± 0.035678).

5 Decide on the appropriate number of significant figures after the uncertainty has been estimated. The examples given in Guideline point **4** illustrate how to make this decision.

6 For repeated measurements, the best estimate for the value of the measured quantity is the mean ($\bar{x}$) of all the measurements (excluding any outliers) and the uncertainty is given by the standard deviation (σ):

$$\sigma = \text{standard deviation} = \sqrt{\frac{\sum_{i=1}^{n}(x_i - \bar{x})^2}{n-1}}$$

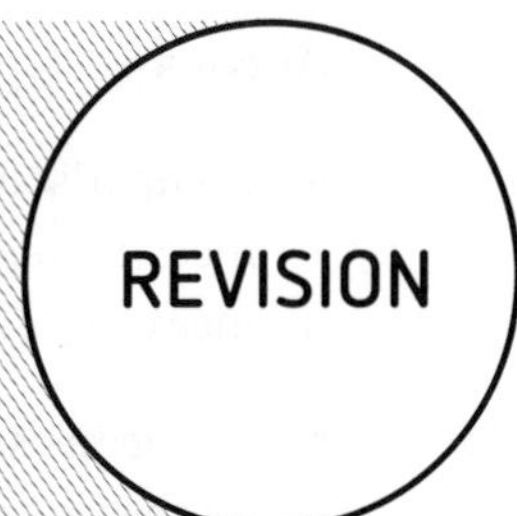

10.1 Important terms

1 Complete the crossword using the clues on page 99.

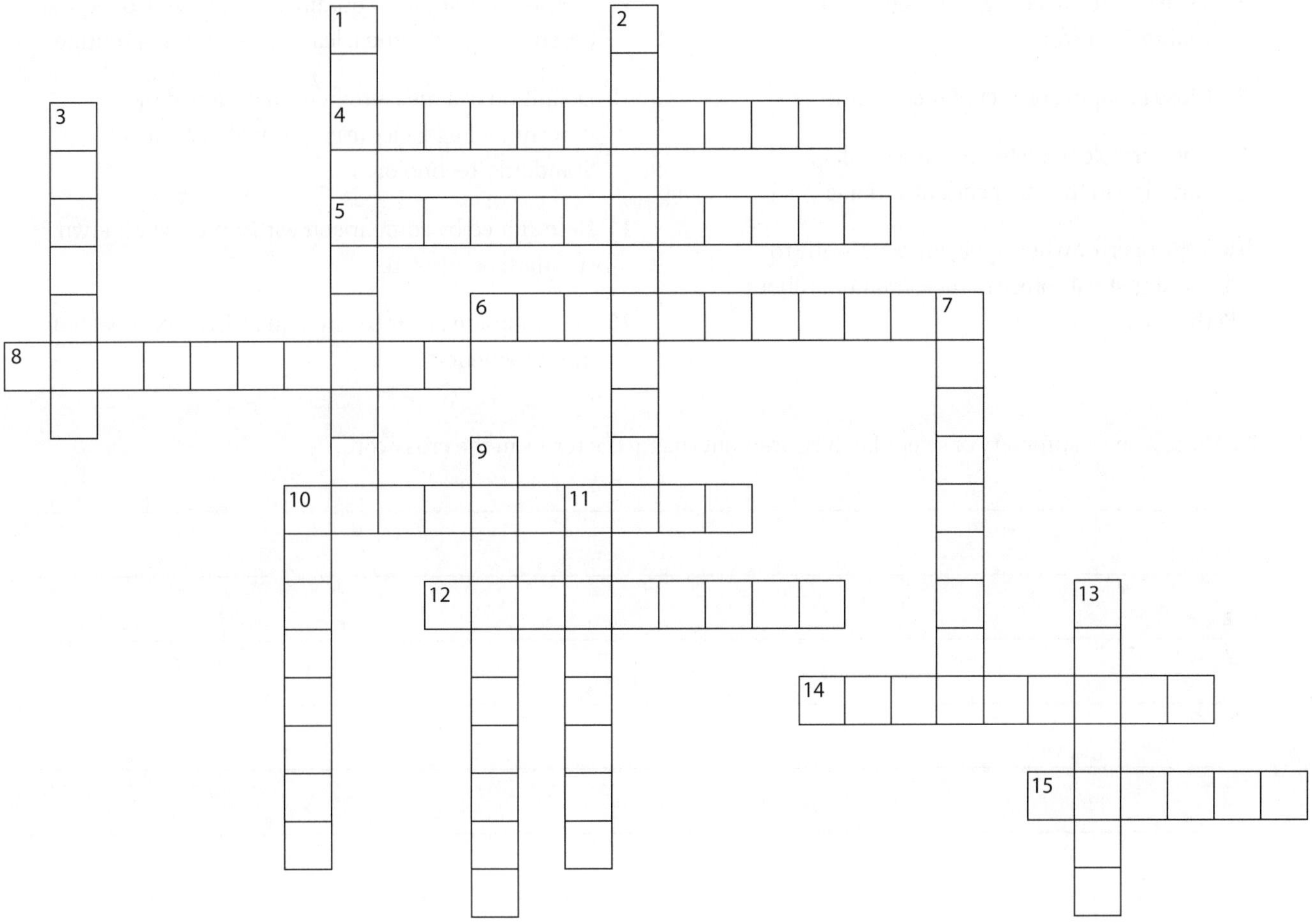

1 2 3 4 5 6 7 8 9 10 11 12 13 14 15

Across

4 Comparison and/or adjustment of device measurement with a known standard of measure

5 Giving the same result to within the established uncertainty

6 A variable upon which another variable depends

8 Limit of reading – the minimum uncertainty in any measurement

10 Type of error which typically occurs due to a calibration error

12 Close grouping of a set of measurements

14 The variable that changes as a result of variation in the independent variable

15 Type of error which typically occurs due to the statistical spread of measurements about the mean

Down

1 An estimate of the range of values within which the 'true value' of a measurement lies

2 Carried out by using a procedure to determine the value (magnitude) of something against a measuring standard

3 Data point that does not fit the pattern displayed by other measured data points

7 An idealisation since all measurements have uncertainty (two words)

9 Method of analysis to obtain a line or curve of best fit based on a model and a least squares fitting routine

10 Definition of a measurement maintained by government agencies (e.g. National Institute of Standards Technology)

11 Determined by comparison with a precisely known standard of measure

13 Best estimate of a measured quantity after repeated measurements

2 Write a short summary of scientific measurements using the terms in the crossword.

9780170412391

10.2 Making measurements, precision versus accuracy

1 State the precision and uncertainty of the following measurements.

a A kitchen timer is set to a time of 20 minutes; it displays 1 minute increments to 60 minutes.

Precision: ____________________ Uncertainty: ____________________

b The room temperature is measured with a mercury thermometer to be 25.6°C.

Precision: ____________________ Uncertainty: ____________________

c The room temperature is measured with a digital thermometer to be 32.9°C.

Precision: ____________________ Uncertainty: ____________________

d An analytical chemist weighs out 1.456g of NaOH using a digital balance.

Precision: ____________________ Uncertainty: ____________________

e The distance between two power poles is measured to be 120m.

Precision: ____________________ Uncertainty: ____________________

f The distance between the H atom and the Cl atom in HCl at equilibrium is given as 0.127nm.

Precision: ____________________ Uncertainty: ____________________

2 At a swimming carnival, the time recorded for the 100m butterfly final was 56.64s.

a What was the precision of the timer used? ____________________

b What was the uncertainty of the timer used? ____________________

c Write down the time as given, including the measurement uncertainty. ____________________

d Given the uncertainty, state the maximum and minimum times that the swimmer could have taken to swim the 100m butterfly race. ____________________

3 In the pairs of measurements in the following table, state the precision and tick the measurement with the lowest uncertainty.

MEASUREMENT	PRECISION	√	MEASUREMENT	PRECISION	√
(1 ± 0.001) nm			1 kilometre ± 0.1 metre		
7 hours ± 15 minutes			1 year ± 1 day		
(25 ± 0.1) mL			(1000 ± 1) mL		
(235.6 ± 0.1) kg			1.364 ± 0.005 g		

10.3 Qualitative and quantitative data

1 Complete the following table by indicating which of the examples describe qualitative data and which describe quantitative data.

DATA	QUALITATIVE OR QUANTITATIVE?
500 mL of ethanol contained in a 1 L beaker	
Describing construction material for a beaker (e.g. glass, plastic)	
Classifying room air (e.g. as dry, humid, warm or cold)	
Stating the mass of a weighed sample as 5.10 g	
The colour of a rock sample	
The surface area per gram of a titanium(IV) oxide (TiO_2) nanosphere = 179.9 $m^2 g^{-1}$	
A rise of 21.2°C in the temperature of a calorimeter	
A molecular weight of 35.5 g	
The three-dimensional geometric shape of molecules (e.g. octahedral, linear or trigonal planar)	
The wavelength of blue light from an argon laser quoted as 488 nm	

10.4 Random and systematic errors

1 In an experiment, the mass of a single 100 mL sample of potassium chloride (KCl) solution is measured by difference. Two *different* precision balances are used to measure the mass. The data obtained are shown in the following table.

MASS BEING MEASURED	BALANCE 1 (g)	BALANCE 2 (g)	AVERAGE MASS WITH UNCERTAINTY
Empty beaker	223.92	227.15	
Beaker + 100 mL KCl solution	325.92	327.15	
100 mL KCl solution (by difference)			

a Complete the table.

b What is the measurement precision in each case? (Add it to the data given in the table.)

c Is the main source of error random or systematic?

____

d What is the procedure used to estimate the uncertainty in the mass of the water itself?

____

____

____

9780170412391

e How would you determine which of the two balances provides the more reliable measurement?

2 In another experiment, the mass of a 100 mL sample of KCl solution (delivered using a measuring cylinder) is again measured by difference. However, the measurement is repeated twice using the same balance. The data obtained are shown in the following table.

	TRIAL 1 (g)	TRIAL 2 (g)	AVERAGE MASS WITH UNCERTAINTY
Empty beaker	223.9	224.4	
Beaker + 100 mL KCl solution	323.6	324.9	
100 mL KCl solution (by difference)			

a Complete the table.

b What is the measurement precision in each case? (Add it to the data given in the table.)

c Is the main source of error random or systematic?

d What is the likely source of the measurement differences?

e How would you determine which of the two balances provides the more reliable measurement?

10.5 Significant figures, absolute uncertainty, percentage uncertainty, percentage error

1 Check the following statements re-expressing quantities in scientific notation. If a statement is correct, write 'Correct' in the space provided. If the answer is incorrect, rewrite the answer correctly.

a $40{,}101{,}000 = 4.0101000 \times 10^7$

b $0.008912 = 8.9 \times 10^{-3}$

c $210{,}708{,}000{,}000 = 2.10708 \times 10^{11}$

2 a Convert the following absolute uncertainties to percentage uncertainties.

i $55.85 \pm 0.02\,\text{mL}$

ii $6.78 \pm 0.02\,\text{cm}$

b Convert the following percentage uncertainties to absolute uncertainties.

i $1.45 \pm 0.01\,\text{L}$

ii $32.33 \pm 0.05\,\text{mm}$

c Calculate the result and state the appropriate uncertainty.

i $(2.95 \pm 0.01)\text{L} + (0.563 \pm 0.005)\text{L}$

ii $(99.99 \pm 0.02)\text{g} - (17.45 \pm 0.01)\text{g}$

iii $(2.95 \pm 0.01)\text{L} \times (0.563 \pm 0.005)\text{L}$

iv $(99.99 \pm 0.02)\text{g} \div (17.45 \pm 0.01)\text{g}$

v $(2.95 \pm 1\%)\text{L} + (0.563 \pm 0.2\%)\text{L}$

9780170412391

vi $(99.99 \pm 0.5\%)\,\text{g} - (17.45 \pm 0.05\%)\,\text{g}$

vii $(2.95 \pm 1\%)\,\text{L} \times (0.563 \pm 0.2\%)\,\text{L}$

viii $(99.99 \pm 0.5\%)\,\text{g} \div (17.45 \pm 0.05\%)\,\text{g}$

ix $2 \times (2.95 \pm 0.01)\,\text{cm}$

x $2 \times (33\,\text{s} \pm 2\%)\,\text{cm}$

xi $(2.95 \pm 0.01\,\text{cm})^2$

xii The Earth's radius is represented by $\frac{4}{3}\pi R^3$ where $R = 6371 \pm 1\,\text{km}$. Do you think the uncertainty given here is appropriate for a statement of the Earth's radius? Explain your answer.

10.6 Presenting and analysing data

Water has a relatively high specific heat capacity, usually quoted as $C = 4.184\,J\,K^{-1}\,g^{-1}$. This means that more heat is required to increase its temperature than is the case for metals (e.g. for copper metal, $C = 0.385\,J\,K^{-1}\,g^{-1}$). This makes water a valuable material for maintaining steady temperatures of devices (e.g. in a car radiator) as a coolant. The high specific heat capacity of water also means that the rate at which the temperature of the atmosphere changes is regulated because of the moisture content. Similarly, the temperature of the ocean remains relatively stable despite daily and seasonal variations in air and land temperature.

One interesting feature of water's heat capacity is that it depends somewhat on temperature. This means that C is not a constant! Table 10.6.1 provides data for the specific heat capacity of water in the temperature range 0–100°C. The last data point is for the heat capacity of water vapour (steam) at 100°C – it is noticeably different from liquid water at the same temperature.

TABLE 10.6.1 The specific heat capacity (C) of water in the temperature range 0–100°C

TEMPERATURE (°C)	SPECIFIC HEAT CAPACITY ($J\,K^{-1}\,g^{-1}$)	TEMPERATURE (°C)	SPECIFIC HEAT CAPACITY ($J\,K^{-1}\,g^{-1}$)
0	4.217	55	4.183
5	4.202	60	4.185
10	4.192	65	4.187
15	4.186	70	4.190
20	4.182	75	4.193
25	4.180	80	4.196
30	4.178	85	4.200
35	4.178	90	4.205
40	4.179	95	4.210
45	4.180	100	4.216
50	4.181	100 (gas)	2.080

1 Which is the independent variable?

____

2 Which is the dependent variable?

____

3 Decide what domain and range you are going to choose for the axes to best display the data.

____

____

4 Plot the dependent variable as a function of the independent variable. Does your choice for the upper and lower limits of the axes ranges make good sense?

5 Use a software program to plot the dependent variable as a function of the independent variable.

6 Determine the temperature at which the specific heat capacity of liquid water is a minimum.

____

7 Employ a suitable regression analysis using a software program to determine a model (e.g. a fourth or sixth order polynomial) for *interpolating* the data to obtain estimates for the specific heat capacity of water at any temperature between 0°C and 100°C. Alternatively, sketch a curve of 'best fit' through the hand-graphed data in Question 4 using the 'eyeball' method.

8 Why is it physically unreasonable to use such a 'curve fit', either via computer analysis or by hand, for *extrapolating* to obtain estimates outside the range of 0–100°C?

____

____

____

____

____

9 Based on your chosen model or method, predict the specific heat capacity of liquid water at 23°C (i.e. interpolate between data points).

____

____

10 Estimate the uncertainty associated with your predicted specific heat capacity at 23°C based on the significant figures provided in Table 10.6.1.

11 Given that the specific heat capacity of water is usually taken to be $4.184\,J\,K^{-1}\,g^{-1}$, use the data in the table to calculate the mean and standard deviation for the measurements and comment on whether the normally quoted value of $4.184\,J\,K^{-1}\,g^{-1}$ is a logical choice.

1 State the number of significant digits in the following measurements.

a 3005 cm

b 65.000 g

c 44.0 mL

d 0.012 s

2 Add or subtract the following measurements and round off your answer to the correct number of significant digits.

a 103.9 g + 0.22 g + 0.175 g

b 105.25 mL − 48.3 mL

3 A solution is made by transferring 10 mL of a 0.1505 ± 0.0003 mol L^{-1} solution, using a 10 mL volumetric pipette (± 0.05 mL) into a (200 ± 0.1) mL volumetric flask. Calculate the final concentration and state the uncertainty.

4 The mass (m) of a sample is estimated by making two measurements: the mass of the sample (A) plus the mass of the weighing vessel; (B) the mass of the empty vessel. Hence, $m = A - B$. Suppose the first measurement is $A = 148.433 \pm 0.002$ and the second measurement is $B = 114.452 \pm 0.002$.

a Calculate m.

b State the uncertainty in the result.

5 A calorimetric experiment involves obtaining the temperature change (ΔT) and then multiplying ΔT by the mass (m) and the heat capacity (C_w) of water to obtain the heat (Q) absorbed by the water in the calorimeter (i.e. $Q = mC_w\Delta T$). Given $\Delta T = 5.2 \pm 0.5°C$, and the mass of water = 100 g, use the normally quoted value for C_w ($C_w = 4.184 \pm 0.005 \, J K^{-1} g$) to determine the value of the heat absorbed. Quote the result to the correct number of significant figures together with an estimate of the uncertainty.

6 A titration experiment with five repeated measurements yields the following values for the concentration (in units of $mol L^{-1}$) of ethanoic (acetic) acid in a sample: 0.1012, 0.0993, 0.1015, 0.0997 and 0.1011. Calculate the mean and the relative standard deviation for this data set.

7 The rate constant for the thermal degradation of a compound measured in the laboratory as a function of temperature yields the data shown in Table 10.7.1.

TABLE 10.7.1 Rate constants for the thermal degradation of a compound as a function of temperature

TEMPERATURE (°C)	RATE CONSTANT (s^{-1})
50	0.350
60	0.790
70	1.270
80	1.458
90	1.997

a Graph the data using appropriate (labelled) scales for the vertical and horizontal axes.

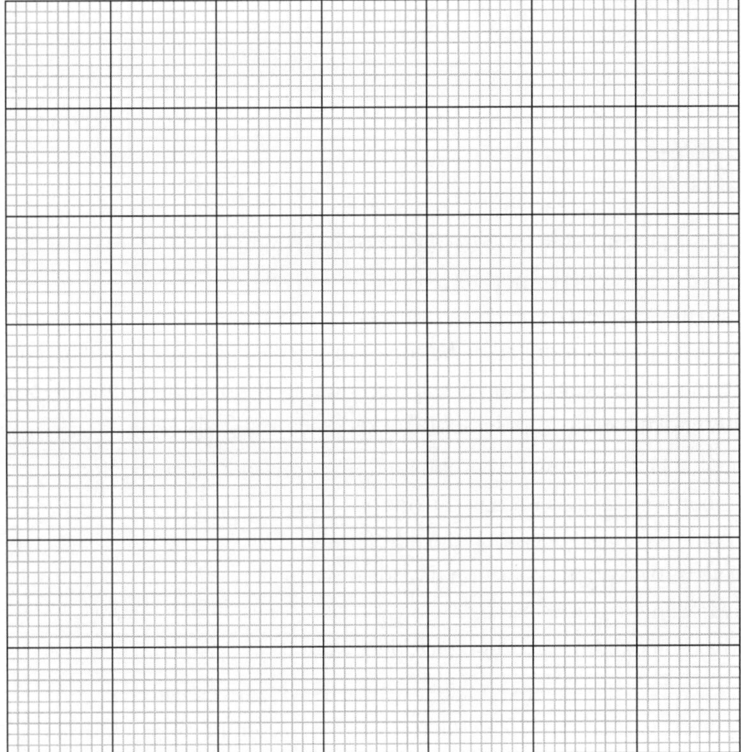

b Use a ruler to estimate a line of best fit for the data.

c Determine the gradient (slope) of the line.

---

d Make an estimate of the uncertainty in the slope based on the precision (number of significant figures) given in the set of data and the spread of the data relative to the straight-line fit.

---

e Quote the result to the correct number of significant figures, including the uncertainty.

---

f Use an extrapolation to estimate the value of the rate constant at 100°C. Quote the result to the correct number of significant figures, including the uncertainty.

---

11 Fuels

LEARNING

Summary

- In Chapter 11, the following classifications were made:
  - fuels as renewable or non-renewable
  - fossil fuels as coal or oil or natural gas
  - nuclear energy produced by fission or fusion
  - alternative fuels as carbon dioxide producing or not carbon dioxide producing
  - alternative fuels as supplemental or replacement fuels.
- In Chapter 11, the following combustion reactions were investigated:
  - Carbon + oxygen → carbon dioxide
  - Octane + oxygen → carbon dioxide + water
  - Methane + oxygen → carbon dioxide + water
  - Hydrogen + oxygen → water
  - Ethanol + oxygen → carbon dioxide + water.
- In Chapter 11, the following nuclear reactions were investigated:
  - Neutron + uranium-235 → telerium-137 + zirconium-97 + neutrons
  - Deuterium + tritium → helium + neutron.
- In Chapter 11, the following topics were discussed in terms of their economic, social and environmental impacts:
  - global warming
  - sea level rising
  - ocean acidification
  - sulfur dioxide inhalation
  - alpha radiation
  - gamma radiation.

9780170412391

11.1 Important terms

1 Complete the following statements by filling in the missing words from the word list below. Research the meaning of any unfamiliar terms from the list using legitimate sources of information.

Word list

alpha radiation	Faraday	magnetic field	sequestration
biomass	fuel	neutron	smog
decomposed	gamma radiation	nuclei	toroidal
density	half-life	organic	transport
domestic	hydrocarbon	performance	vapour
Edison	impurity	photolysis	viscosity
efficiency	industrial	plasma	volatility
electricity	infrared radiation	poloidal	
electromagnetic induction	kinetic	power	

a The ______________ of a substance is its mass divided by its volume.

b The ______________ of a liquid is its resistance to flow.

c The ______________ is the tendency of a substance to vaporise.

d ______________ is the gaseous phase of water.

e ______________ is emission of a helium nucleus.

f ______________ is short wavelength, high energy radiation that will pass through flesh and bone.

g ______________ is invisible to the human eye, emitted by all objects and sensed as heat.

h A ______________ is burnt to produce heat or ______________.

i There are three major sectors that use fuels: ______________, ______________ and ______________.

j Fossil fuels are produced from ______________ matter that has ______________ over a long time under high pressure and temperature.

k Sulfur is an ______________ found in fossil fuels.

l No fuel has 100% ______________.

m A ______________ contains a sea of charged particles because electrons are free to move independently of the nucleus.

n A ______________ compound contains only carbon and hydrogen atoms.

o Nuclear fusion is joining of small ______________.

p The optimal ratio of energy outputs to energy inputs for a nuclear fusion power plant is known as its ________________.

q A ________________ field is generated by wrapping coils around the poles of a sphere or donut shape.

r A ________________ field is generated by wrapping coils around the lines of latitude of a sphere or donut shape.

s There are two types of ________________: traditional (London) and photochemical (Los Angeles).

t ________________ is sourced from living or recently living plant and animal materials.

u A ________________ has about the same mass as a proton, but without any charge.

v ________________ is the time required for a quantity of a substance to be reduced to half its initial value.

w Thomas ________________ generated ________________ energy to drive a steam train.

x Michael ________________ used the ________________ of a magnet to produce ________________ in a coil of wire via ________________.

y ________________ is a chemical reaction occurring by the action of light.

z Carbon ________________ is the process involved in the capture and storage of atmospheric carbon dioxide.

2 Match each term with the appropriate description using an arrow (or list letter pairs that go together).

Term	Description
a Biofuels	**A** Decomposed organic matter under high temperature and pressure for a long time
b Combustion	**B** Unable to be regenerated at a similar rate to production time
c Fossil fuels	**C** Rapid chemical reaction of a substance with oxygen producing energy
d Greenhouse effect	**D** Gas that traps outgoing infrared radiation
e Greenhouse gas	**E** Trapping of outgoing infrared radiation heating Earth's near-surface atmosphere
f Non-renewable	**F** Splitting of heavy nuclei
g Nuclear fission	**G** Can be replenished and is essentially limitless
h Nuclear fusion	**H** Fuels that are primarily made of biomass
i Renewable	**I** Joining of small nuclei.

3 Complete the following table comparing the properties of fuels.

FUEL	EQUATION	ADVANTAGES	DISADVANTAGES
Coal		Tried and tested	
		Less sulfur and nitrogen oxides and dioxides	More unburnt hydrocarbons
		Less unburnt hydrocarbons	More sulfur and nitrogen oxides and dioxides
Natural gas			Potential explosions of wells and pipelines
Hydrogen			Low density of hydrogen
	${}^{1}_{0}n + {}^{235}_{92}U \rightarrow {}^{137}_{52}Te + {}^{97}_{40}Zr + 2{}^{1}_{0}n$		Extraction of fuel, storage of waste, reactor explosions
	${}^{2}_{1}H + {}^{3}_{1}H \rightarrow {}^{4}_{2}He + {}^{1}_{0}n$		Currently low performance
Biofuel		Renewable	

4 Complete the following crossword. Some clues have two-word answers, but do not insert a space between the two words of your answer. Some answers are chemical formulas; for example, if the answer was H_2O, write as 'HTWOO' into the grid.

Across	**Down**
1 Ethanol is a type of ________	**2** Produced from decomposed organic matter
3 Clean fuel whose low density hinders its widespread use	**3** Product of nuclear fusion
5 Splitting of heavy nuclei	**4** Additive in E10 fuel
6 Fuel for nuclear fission	**9** Fossil fuels are ____________
7 Source of tritium	**11** Part of nuclear power plant that exploded at Chernobyl
8 Hydroxide used in carbon sequestration	**13** A coil turning in the presence of a magnetic field generates ________
10 Type of power plant at Fukushima	**15** Composed of methane
12 Type of oxides and dioxides produced by diesel	**16** Process fuels undergo to produce energy
14 Joining of small nuclei	**17** Effect of heating the surface temperature of the Earth
18 Wind power is ______________	
19 Dioxide responsible for deaths during The Great Smog of 1952	
20 Produces less unburnt hydrocarbons than petrol	
21 Produced by gasification of wood	

5 Write a short summary about fuels using the terms in the crossword.

11.2 Earth-like planet surface temperatures

The data in Table 11.3.1 uses the equations shown to calculate the effective surface temperatures of Venus, Earth and Mars (°C).

Mathematical equation

$$T_e(\text{K}) = \left(\frac{[(1-\text{A})F_s]}{4\sigma}\right)^{1/4}$$

Word equation

Effective surface temperature in Kelvin = {[(one minus albedo) times the Sun's flux] divided by four times Boltzmann's constant} raised to the power of one-quarter

where $\sigma = 5.68 \times 10^{-8}\,\text{W}\,\text{m}^{-2}\text{K}^{-4}$ and $0°\text{C} = 273.15\,\text{K}$.

TABLE 11.3.1 Characteristics of Mars, Earth and Venus

PLANET	MARS	EARTH	VENUS
Albedo (A)	0.14	0.31	0.77
Sun's flux (F_s, $\text{W}\,\text{m}^{-2}$)	593	1368	2643
Atmospheric pressure (kPa)	0.6	101	9400
Percentage composition of carbon dioxide (%)	95.3	0.03	96
Actual temperature (°C)	–50	+17	+232

9780170412391

1 Apply your knowledge of the greenhouse effect to compare its consequences with the actual surface temperatures of the planets shown in Table 11.3.1.

2 Would it be theoretically possible for humans to inhabit either of the other Earth-like planets? Explain your answer in terms of the human requirements to breathe and maintain a body temperature of 37°C.

11.3 Snowfall acidification

1 Apply your knowledge of carbon dioxide reacting with water to determine why the pH of a snowfall in Antarctica is 5.82 and the pH of a snowfall in Scotland is 3.55.

2 Apply your knowledge of renewable energies to determine how the pH of the Scottish snowfall (3.55) could theoretically be increased closer to that of the Antarctica snowfall (5.82).

3 Discuss the social and economic implications of making the theoretical changes suggested in question **2**.

1 Which one of the following would not be considered a biofuel?

A green gasoline

B biodiesel

C diesel

D biogas

2 Which of the equations below represents the combustion of diesel fuel?

A $CH_4 + 2O_2 \rightarrow CO_2 + 2H_2O$

B $2C_8H_{18} + 25O_2 \rightarrow 16CO_2 + 18H_2O$

C $C_2H_5OH + 3O_2 \rightarrow 2CO_2 + 3H_2O$

D $2C_{16}H_{34} + 49O_2 \rightarrow 32CO_2 + 34H_2O$

3 When two light nuclei combine to form a heavier nucleus producing large amounts of heat, this process is referred to as:

A Nuclear power

B Nuclear fusion

C Nuclear fission

D Nuclear transmutation

4 Which of the following is **not** considered to be a greenhouse gas?

A $H_2O(g)$, water vapour

B $CH_4(g)$, methane

C $CO_2(g)$, carbon dioxide

D $N_2(g)$, nitrogen

5 Natural gas is an example of a _______________ energy resource.

6 Sugar cane crops produce which additive to petrol?

____

7 Natural gas produces relatively clean, low emissions energy. What property of this fuel prevents its widespread use?

____

8 Nuclear fusion is a clean, renewable energy source capable of producing large amounts of energy. What aspect of the production of this energy prevents its widespread use?

____

9 With the aid of equations, explain why burning fossil fuels such as coal and oil gives rise to acid rain.

____

____

____

10 Pure hydrogen gas can be produced on an industrial scale by steam methane reforming. With the aid of equations, explain how this process works and suggest one environmental problem with this process.

____

____

____

____

11 Below is a bar graph showing the energy density of a number of fuels along with their relative amount of carbon dioxide production.

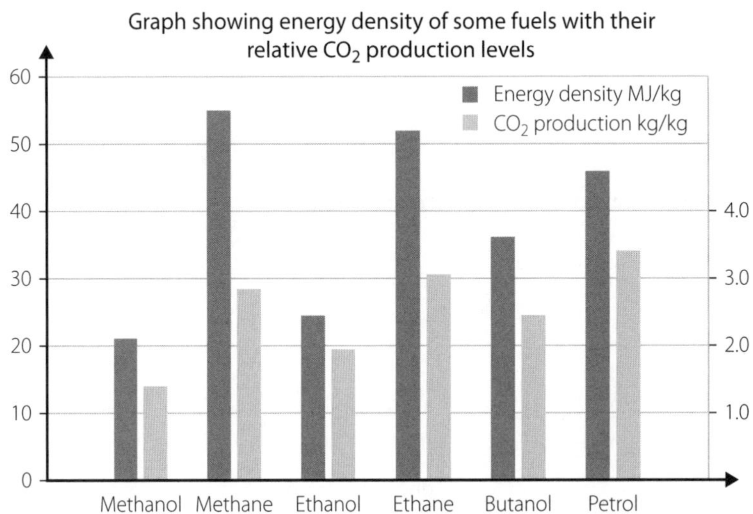

From this graph, it can be seen that methanol has an energy density of about 21 megajoules of energy (21×10^6 J) per kg of methanol. However, it produces about 1.4 kg of carbon dioxide for every kg of methanol burnt.

9780170412391

a From this graph, evaluate each fuel in terms of its energy density and CO_2 production. Illustrate your answer with balanced chemical equations for the combustion of each fuel.

b Discuss what this data means in terms of using biofuels as a viable, long-term alternative to using fossil fuels to supply our energy needs.

12 Mole concept and the law of conservation of mass

Summary

- The calculation of chemical quantities requires the use of measurements related to atoms and molecules. The relative atomic mass of an element is defined as the ratio of the weighted average mass per atom of the naturally occurring form of the element to one-twelfth the mass of an atom of carbon-12.
- Relative atomic masses reflect the isotopic composition of the element.
- The Avogadro constant (N_A) is the number of atoms (6.02×10^{23}) in exactly 12 g of the carbon-12 isotope.
- A mole is the amount of substance containing 6.02×10^{23} particles of that substance.
- The relationship between the number of moles (n) of a substance and the number of particles (atoms, ions or molecules) is given by:

$$n = \frac{\text{Number of particles}}{6.02 \times 10^{23}}$$

- Law of conservation of mass: mass is neither created nor destroyed by chemical or nuclear reactions, radioactive decay or physical transformations.
- The relationship between the amounts of reactants and products in a chemical reaction is called 'stoichiometry'.
- The relative molecular mass of a substance (M_r) is the mass of one molecule of the substance, calculated by adding the relative atomic masses of all the component atoms of the molecule, on a scale in which the mass of an atom of the carbon-12 isotope is exactly 12. For an atom, the relative atomic mass is denoted A_r.
- The relative formula mass for an ionic substance (also given the symbol M_r), is the mass of one formula unit of the ionic compound also on a scale in which the mass of an atom of the carbon-12 isotope is exactly 12.
- The percentage composition of a compound is the percentage by mass of each of the different elements in the compound.
- One mole of any substance has a mass equal to the relative atomic, molecular or formula mass of that substance expressed in grams. The mass of one mole of a substance is termed the molar mass (M).
- The relationship between the number of moles (n), mass (m) and the molar mass (M) of a substance is given by:

$$n = \frac{m}{M}$$

9780170412391

- The empirical formula of any compound is the simplest whole number ratio in which the atoms of the elements are present.
- In a chemical reaction, the limiting reagent is the reactant that determines the amount of products that can be produced. The other reactants are referred to as being in excess.
- $\text{Percentage}(\%)\ \text{yield} = \frac{\text{experimental yield}}{\text{theoretical yield}} \times \frac{100}{1}$
- The mass relationships between reactants and products are derived from the mole ratios in a balanced chemical equation.
- To determine the theoretical yield of a product formed in a chemical reaction, starting with a given mass of reactants,
  (i) use the mole ratios in a balanced chemical equation for the reaction
  (ii) make use of the relationship that one mole of any substance has a mass equal to the relative atomic, molecular or formula mass of that substance expressed in grams.

REVISION

12.1 Important terms

1 Match each term with the appropriate description using an arrow (or list letter pairs that go together).

Term	Description
a Anions	A Amount of substance containing 6.02×10^{23} particles
b Atomic mass unit	B Compound consisting of oppositely charged ions held in a lattice
c Species	C Consists of different types of atoms covalently bonded
d Avogadro constant (N_A)	D Atoms, ions or molecules involved in a chemical reaction
e Balanced equation	E Empirical formula of any ionic compound or network solid
f Cations	F Formula of an element or compound
g Chemical equation	G Gives actual number of different atoms present in a molecule
h Chemical formula	H Group of atoms held together by covalent bonds
i Empirical formula	I Mass of an atom compared to the mass of a carbon-12 atom
j Formula unit	J Mass of one mole of a substance
k Ionic compound	K Mass of a molecule compared to the mass of a carbon-12 atom
l Lattice	L Negatively charged ions
m Mass (*m*)	M New substances formed in a chemical reaction
n Molar mass (*M*)	N Number of atoms in exactly 1 g of the carbon-12 isotope
o Mole of a substance	O Numbers of each type of atom are the same on both sides
p Molecular compound	P Percentage by mass of each element found in a compound
q Molecular formula	Q Positively charged ions
r Molecule	R Quantity of matter in a substance
s Percentage composition	S $\frac{1}{12}$ the mass of one carbon-12 atom
t Products	T Regular arrangement of particles within a solid
u Reactants	U Mass of a formula unit compared to the mass of a carbon-12 atom
v Relative atomic mass (A_r)	V Simplest whole number ratio of atoms in a compound
w Relative formula mass (M_r)	W Starting substances in a chemical reaction
x Relative molecular mass (M_r)	X Study of the amounts of substances in a chemical reaction
y Stoichiometry	Y Uses chemical symbols for reactants and products in a reaction

12.2 Mole and Avogadro's number of particles

1 Use the following definitions and relationships to complete the table. (You will also need the refer to the periodic table of elements on page 3).

Avogadro's number (N_A): 6.022×10^{23}

Molar mass (M): atomic mass (m) or sum of atomic masses for each atom composing the molecule, stated in grams (e.g. for H_2O, $M = 2 \times 1.008 + 1 \times 16 = 18.016$ g)

Number of moles: n

Number of particles (i.e. either atoms or molecules) in one mole: $n \times N_A$

Number of moles of particles in a mass (m) of either atoms or molecules): $n = \frac{m}{M}$

Mass of n moles in grams: $m = n \times M$

SUBSTANCE	MOLAR MASS (g) (M)	NUMBER OF PARTICLES ($n \times N_A$)	NUMBER OF MOLES OF PARTICLES (n)	MASS OF n MOLES (g)
Germanium (Ge)			2.5 Moles of germanium atoms	
Methanoic (formic) acid (HCOOH)				138 g of methanoic acid
Calcium phosphate ($Ca_3(PO_4)_2$)		2.409×10^{24} Molecules of calcium phosphate		
Xenon hexafluoride (XeF_6)			3.0 Moles of xenon hexafluoride molecules	
Dichloromethane (CH_2Cl_2)				255 g of dichloromethane

2 How many atoms are there in one mole of atoms?

____

3 How many atoms are there in 28.09 g of silicon atoms?

____

4 What is the mass of one mole of Cl atoms?

____

5 How many molecules are there in one mole of oxygen molecules (O_2)?

____

6 How many molecules are there in 159.8 g of Br_2 molecules?

____

7 How many molecules are there in a teaspoon containing 6.3 g of sodium fluoride (NaF)?

____

8 What is the mass of one mole of Cl_2 molecules?

____

9 What is the mass of one mole of water (H_2O)?

____

10 What is the molar mass of potassium dichromate ($K_2Cr_2O_7$)?

____

11 What is the molar mass of sucrose ($C_{12}H_{22}O_{11}$)?

____

12 Calcium carbonate has the formula $CaCO_3$. In 2 moles of $CaCO_3$, state the number of:

a moles of Ca^{2+} ions

____

____

b moles of CO_3^{2-} ions

____

____

c moles of oxygen atoms

____

____

d individual atoms of oxygen.

____

____

12.3 Empirical and molecular formulas

WORKED EXAMPLE

1 A 100.00 g sample of tetraethyl lead, a gasoline additive, is found to contain 64.06 g of lead (Pb), 29.71 g of carbon (C), and 6.23 g of hydrogen (H). Use these data to determine the empirical formula of tetraethyl lead.

ANSWER

a Calculate the relative number of moles (mole fractions) of Pb, C and H in the sample (molar masses (g): Pb, 207.2; C, 12.01; H, 1.008).

$$\text{Mole fraction Pb} = \frac{64.06}{207.2} = 0.186$$

$$\text{Mole fraction C} = \frac{29.71}{12.01} = 1.484$$

$$\text{Mole fraction H} = \frac{6.23}{1.008} = 3.710$$

Therefore, the ratio of the number of moles of Pb:C:H present in the sample = 0.186 : 1.484 : 3.710.

b Using 0.186 (the smallest mole fraction) as a divisor, we obtain the empirical formulas.

$$Pb:C:H = \frac{0.186}{0.186}:\frac{1.484}{0.186}:\frac{3.710}{0.186} = 1:8:20$$

Therefore, the empirical formula of tetraethyl lead is PbC_8H_{20}.

9780170412391

QUESTIONS

1 A 170.00 g sample of an unidentified compound contains 29.84 g of sodium, 67.49 g of chromium, and 72.67 g of oxygen. What is the compound's empirical formula?

2 An oxide of chromium is found to have the following percentage composition: 68.4% Cr and 31.6% O. Determine the compound's empirical formula.

3 Calculate the percentage composition of NH_4OH.

4 The empirical formula of a hydrocarbon is found to be CH_2. Laboratory procedures have found that the molar mass of the compound is 84.16 g mol^{-1}. What is the molecular formula of this compound?

5 Phenyl magnesium bromide is used as a reagent in organic synthesis. Its molar mass is known to be $181.3\,g\,mol^{-1}$. It contains 39.75% C, 2.79% H, 13.41% Mg and 44.07% Br. Determine its empirical and molecular formula.

______

______

______

______

______

12.4 Percentage yield and theoretical yield

WORKED EXAMPLE

1 How many grams of hydrochloric acid (HCl) are required to react completely with 25 g of calcium hydroxide ($Ca(OH)_2$) in the following reaction? If 31.50 g of calcium chloride ($CaCl_2$) is collected in the experiment, calculate the experimental percentage yield of calcium chloride.

$$2HCl + Ca(OH)_2 \rightarrow CaCl_2 + 2H_2O$$

ANSWER

a Calculate the theoretical yield of calcium chloride.

$$\text{Number of moles } (n) \text{ of } Ca(OH)_2 = \frac{25}{Ca(OH)_2 \text{ molar mass}} = \frac{25}{74.10} = 0.34 \text{ moles}$$

Number of moles of HCl required $= 2 \times 0.34 = 0.68$ moles

Therefore, the mass of HCl required $= 0.68$ moles $\times$ HCl molar mass $= 0.68 \times 36.46\,g = 24.8\,g$.

b Calculate the theoretical yield of calcium chloride.

1 mole of $Ca(OH)_2$ yields 1 mole of $CaCl_2$.

Therefore, the theoretical yield of $CaCl_2 = 0.34$ moles $\times$ $CaCl_2$ molar mass $= 0.34 \times 111.0\,g = 37.7\,g$.

c Calculate the percentage yield of calcium chloride.

$$\text{Percentage}(\%)\text{ yield} = \frac{\text{experimental yield}}{\text{theoretical yield}} \times 100\% = \frac{31.50}{37.44} = 84.1\%$$

Therefore, the percentage yield is 84.1%.

QUESTIONS

1 The combustion of 64.0 g of oxygen (O_2) with 46.0 g of ethanol (C_2H_5OH) takes place according to the following equation.

$$C_2H_5OH(l) + O_2(g) \rightarrow CO_2(g) + H_2O(l)$$

a Is the equation balanced? If not, balance it.

______

b What is the limiting reactant?

______

c What is the theoretical yield of carbon dioxide (CO_2)?

2 Consider the reaction of lithium hydroxide (LiOH) and potassium chloride (KCl).

$$LiOH + KCl \rightarrow LiCl + KOH$$

a If the reaction is carried out using 20.0 g of lithium hydroxide, what is the theoretical yield of lithium chloride (LiCl)?

b If 6 g of lithium chloride were produced in the experiment, calculate the percentage yield.

3 Consider the following reaction of phosphoric acid (H_3PO_4) with potassium hydroxide (KOH).

$$H_3PO_4 + KOH \rightarrow K_3PO_4 + H_2O$$

a Balance the equation.

____

b If 50.0 g of phosphoric acid is reacted with excess potassium hydroxide, determine the theoretical yield.

____

____

____

____

____

____

____

c Determine the percentage yield of potassium phosphate (K_3PO_4) if 63.0 g of potassium phosphate are isolated in the experiment.

____

____

____

____

____

____

____

____

9780170412391

1 Decide whether each of the following statements is True or False. Circle the correct answer. If false, provide the correct statement.

a The relative atomic mass of an atom is its mass relative to any carbon atom. (True / False)

____

b The relative molecular mass of a molecule is the sum of the relative atomic masses of the atoms in that molecule. (True / False)

____

c All pure substances exist as molecules. (True / False)

____

d The formula of an ionic compound gives the ratio of the atoms present in the compound. (True / False)

____

e The Avogadro constant is equal to the number of atoms in exactly 12 g of the carbon-12 isotope. (True / False)

____

f One mole of a substance would contain 6.022×10^{-23} particles of that substance. (True / False)

____

g The formula of any ionic substance is always an empirical formula. (True / False)

____

h Molar mass is an actual mass that is measured in moles. (True / False)

____

i A molecular formula gives the exact number of the different kinds of atoms in the molecule. (True / False)

____

j Stoichiometry focuses on the actual amounts of reactants and products in a reaction. (True / False)

____

2 Calculate the relative molecular masses (M_r) of the following molecules, given the following data (where A_r is the relative atomic mass): $A_r(C) = 12.01$, $A_r(H) = 1.01$, $A_r(N) = 14.01$, $A_r(O) = 16.00$, $A_r(S) = 32.06$.

a Ammonia (NH_3); $M_r(NH_3) =$

b Sulfuric acid (H_2SO_4); $M_r(H_2SO_4) =$

c Carbonic acid (H_2CO_3); $M_r(H_2CO_3) =$

3 Calculate the relative molecular masses (M_r) of the following molecules, given the following data (where A_r is the relative atomic mass): $A_r(H) = 1.01$, $A_r(Mg) = 24.31$, $A_r(N) = 14.01$, $A_r(Na) = 22.99$, $A_r(O) = 16.00$, $A_r(S) = 32.06$.

a Sodium sulfate (Na_2SO_4); $M_r(Na_2SO_4) =$

b Magnesium nitrate ($Mg(NO_3)_2$); $M_r(Mg(NO_3)_2) =$

c Magnesium hydroxide ($Mg(OH)_2$); $M_r(Mg(OH)_2) =$

4 If a person breathed in 3011×10^{18} molecules of nitrogen dioxide (NO_2), a common air pollutant in urban environments, how many grams of nitrogen dioxide would have been inhaled?

5 a Balance the following equation, which shows the formation of Cisplatin (diamminedichloroplatinum(II)), a drug used in chemotherapy, from the precursor chemical, potassium tetrachloroplatinate(II) (K_2PtCl_4).

$$K_2PtCl_4 + NH_3 \rightarrow Pt(NH_3)_2Cl_2 + KCl$$

b Determine the theoretical yield of potassium chloride (KCl) if you start with 56 g of ammonia (NH_3) and an excess of potassium tetrachloroplatinate(II).

____

____

____

____

c Which is the limiting reagent given these conditions?

____

d Determine the experimental percentage yield of $Pt(NH_3)_2Cl_2$ if 2.5 g of $Pt(NH_3)_2Cl_2$ are collected in the reaction.

____

____

____

____

____

UNIT TWO

MOLECULAR INTERACTIONS AND REACTIONS

- Topic 1: Intermolecular forces and gases
- Topic 2: Aqueous solutions and acidity
- Topic 3: Rates of chemical reactions

13 Intermolecular forces

LEARNING

Summary

- The forces that exist between molecules in a substance are governed, in part, by the shape of the molecules.
- The shapes of molecules can be explained and predicted using the valence shell electron pair repulsion theory (VSEPR).
- The VSEPR theory suggests that if the valence (outer) electrons of the central atom in a simple molecule are arranged in pairs– bonding pairs and non-bonding (lone) pairs, the number of pairs of electrons determines the shape of the molecule.
- The forces that exist between molecules in a substance are governed, in part, by the presence of polar bonds within the molecule.
- A polar bond is caused by the unequal sharing of electrons in the bond due to a difference in electronegativity of the atoms in the bond.
- A bond is considered non-polar covalent if the electronegativity difference is less than 0.5, polar covalent if it is between 0.5 and 1.9 and ionic if it is greater than 2.0.
- A molecule can contain polar bonds yet be non-polar overall due to the shape of the molecule.
- Intermolecular forces include dispersion forces, dipole–dipole interactions and hydrogen bonding.
- Dispersion forces are the weakest of the intermolecular forces and are caused by the temporary attraction of the nucleus of an atom and the electron cloud of a neighbouring atom. This means that all substances exhibit dispersion forces to some degree.
- Dipole–dipole interactions are caused when polar molecules interact.
- Hydrogen bonding, the strongest intermolecular force, is a type of dipole–dipole interaction occurring between molecules that contain at least one H atom bonded to an F atom, an O atom or an N atom (H—FON).
- Intermolecular forces affect the physical properties of covalent substances such as melting and boiling points, solubility and vapour pressure.

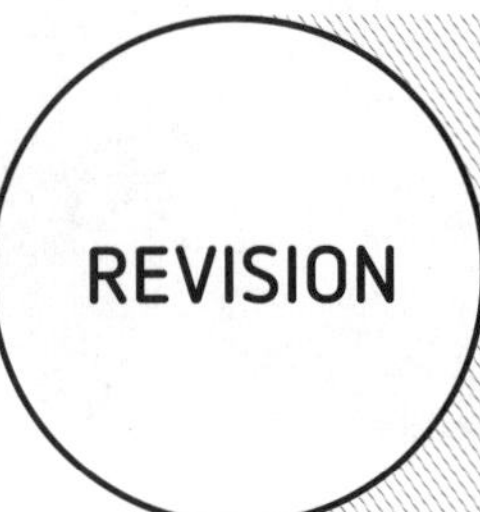

13.1 Important terms

An incomplete concept map for intermolecular forces is provided with some of the concepts filled in.

1 Use the following concept list to complete the concept map.

Concept list

dispersion	electronegativity	hydrogen bonding	melting/boiling point
polar molecule	solubility	vapour pressure	

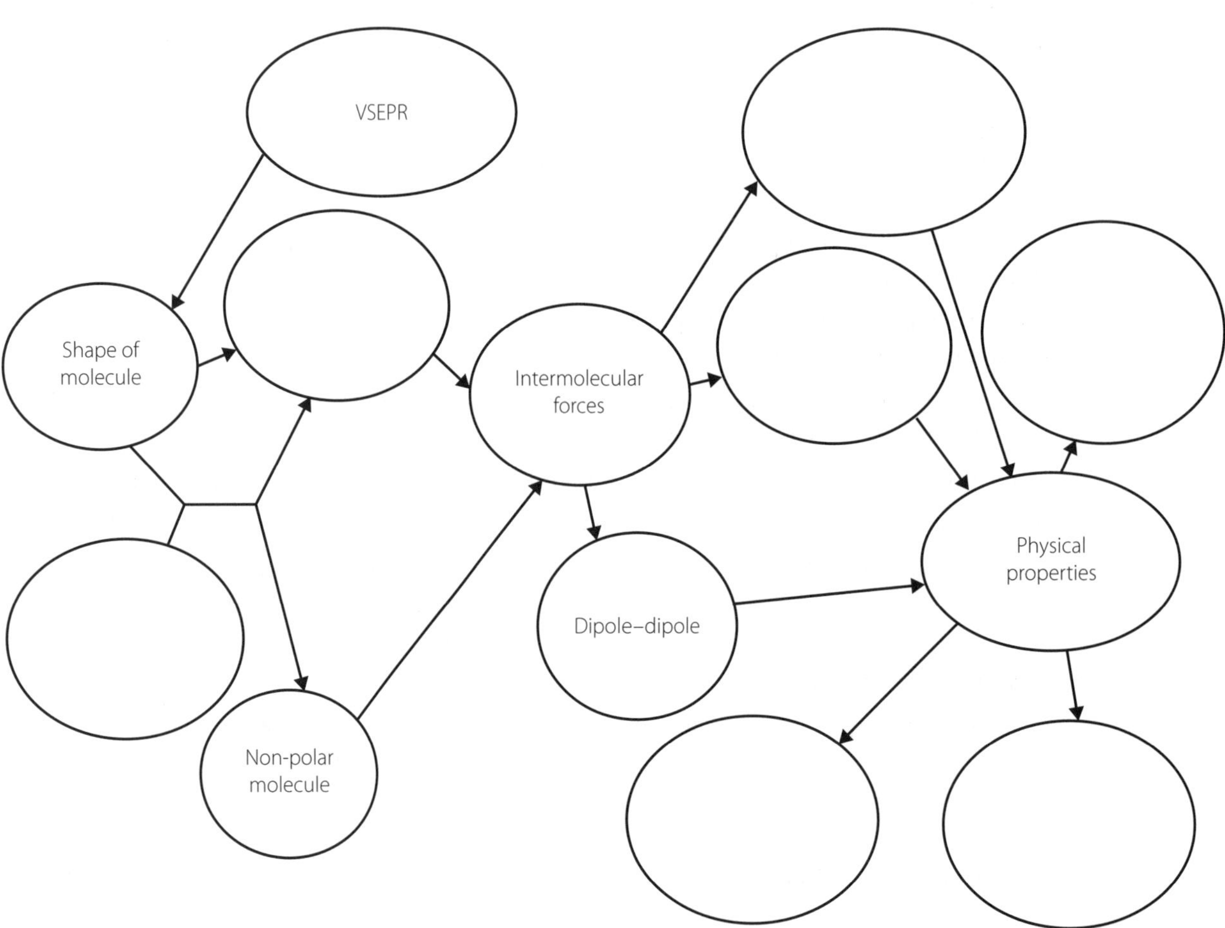

13.2 Valence shell electron pair repulsion (VSEPR) theory

WORKED EXAMPLE

1 Nitrogen (N) and phosphorus (P) are in the same group in the periodic table. The boiling point of ammonia (NH_3) is –33.4°C whereas the boiling point of phosphine (PH_3) is –87.7°C. Explain this difference using VSEPR.

ANSWER

a Draw the electron-dot formula for NH_3 as shown in Figure 13.2.1.

FIGURE 13.2.1 Electron dot formula for ammonia

H: N̈ :H
    Ḧ

b Count the number of electron pairs around the central atom.

- The central atom (N) has four electron pairs around it and so forms a tetrahedron:
- 3 bonding pairs of electrons – 3 N — H bonds
- 1 lone pair of electrons

FIGURE 13.2.2 The tetrahedral shape of ammonia

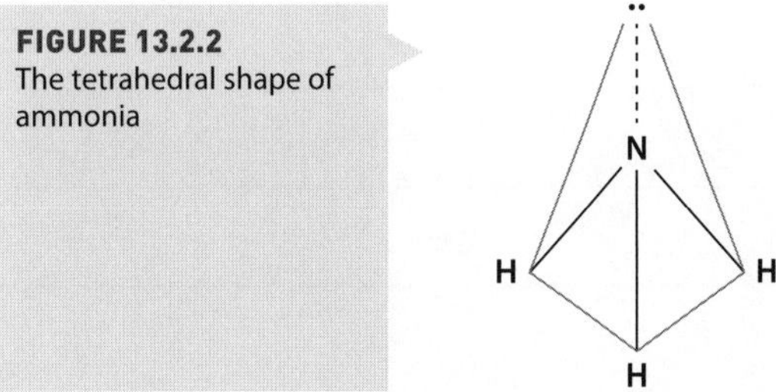

c Check for lone pairs.

Lone pairs help to determine the shape of a molecule but they are never part of the resulting shape. So, instead of the lone pair forming the top corner of the shape, it is the N atom that does this. The shape is called a trigonal pyramid.

d Check for polar bonds and molecular symmetry.

The N—H bond is polar because the difference in electronegativity between the atoms is:

$$3.0\,(N) - 2.2\,(H) = 0.8$$

The molecular symmetry of the molecule (trigonal pyramidal) means that each N—H dipole is pointing to the N atom and so the molecule is **polar**.

e Determine the intermolecular forces.

Because NH_3 is a polar molecule it would be reasonable to expect it to exhibit dipole–dipole interactions.

Remember that hydrogen bonding is a special kind of dipole–dipole interaction and so the molecule needs to be checked for H—FON (hydrogen bonded to fluorine, oxygen or nitrogen). NH_3 has three hydrogen atoms bonded to nitrogen and therefore exhibits hydrogen bonding.

f Determine the dominant intermolecular force

If steps **a–e** are carried out for PH_3, you will see that the dominant intermolecular force in this molecule is dipole–dipole interactions. These are not as strong as hydrogen bonding and this accounts for ammonia having a higher boiling point than phosphine. The molecules in ammonia are held together more strongly and so more heat needs to be applied to overcome these forces.

QUESTIONS

The shapes of simple molecules can be predicted using the VSEPR theory. This theory considers the number of pairs of electrons around the central atom in a molecule. These pairs can be bonding or non-bonding (lone) pairs. The number of electron pairs determines the geometrical shape of the molecule, which can then be modified by the presence of lone pairs.

1 Use VSEPR theory to predict the shape of each of the following substances.

a H_2S

____

____

____

b BH_3

____

____

____

c HCl

____

____

____

d SiH_4

____

____

____

e SO_2

____

____

____

f NO_2

____

____

____

2 Fill in the gaps in the following sentences using the word list given below. (Hint: the same word can be used more than once.)

Word list

bent	bent	lone pairs	MX_2E_2
MX_3E	MX_3E	tetrahedral	tetrahedral
trigonal planar	trigonal planar	trigonal planar	trigonal pyramid

a Water (H_2O) has a __________ arrangement of electron pairs around the central O atom but its actual shape is __________ due to the presence of ________________ around the O atom.

b Ammonia (NH_3) has a ______________ arrangement of electron pairs around the central N atom, whereas boron trihydride (BH_3) has a __________________ arrangement of electron pairs around the central B atom.

This explains why NH_3 is __________________ shaped and BH_3 is ________________ shaped.

c If a molecule had a general formula of MX_3 it would be ______________ shaped but if it had the formula MX_2E it would be ________________, even though it contained the same number of electron pairs around the central atom. Likewise, a tetrahedral arrangement could be produced by a MX_4, a ____________, or a __________________.

13.3 Polarity of molecules

In covalent bonding, the electrons are shared by the atoms that make up the bond. However, the electrons are not necessarily shared evenly between the atoms. Some elements have a greater affinity or attraction for electrons than others. The degree to which elements attract electrons is called 'electronegativity' (see Table 13.3.1).

The degree of electronegativity difference between atoms in a covalent bond determines the *polarity* of the bond, causing *dipoles*.

TABLE 13.3.1 Electronegativity values of some elements

ELEMENT	ELECTRONEGATIVITY
H	2.2
He	–
Li	1.0
C	2.5
N	3.0
P	2.2
O	3.5
S	2.6
F	4.0
Ne	–
Cl	3.0
Br	2.8
Ar	–

1 Use the information in Table 13.3.1 to rank the covalent bonds in the following table in order of decreasing polarity.

BOND	RANK
H—Cl	
P—H	
N—H	
C—O	
S—F	
O—H	
F—F	
C—Br	
C—F	

13.4 Observable properties and intermolecular forces

Molecular geometry and bond polarity determine the type of intermolecular forces that occur between substances. These forces are classified as dispersion, dipole–dipole interactions and hydrogen bonding.

An understanding of how intermolecular forces work is important when explaining physical properties of substances such as melting point, boiling point, solubility in polar and non-polar substances and vapour pressure.

1 Complete the intermolecular forces box and the shape box in the following diagram. Then connect each molecule to the appropriate intermolecular force and shape using arrows, as shown for CH_4.

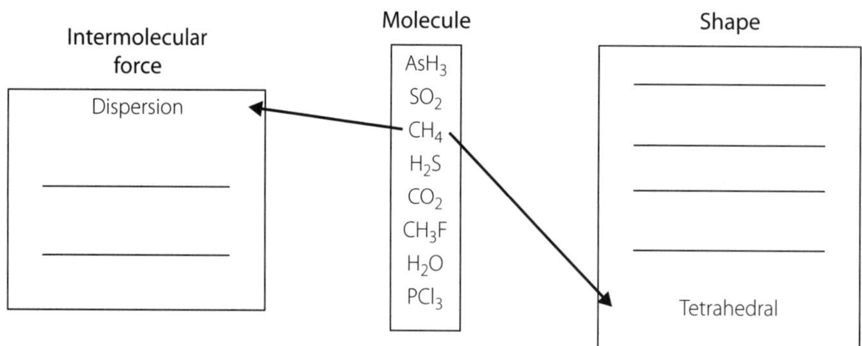

For questions **1** to **3**, circle the best option.

1 Which of the following compounds should have the lowest boiling point?

A HCl

B PH_3

C H_2O

D SiH_4

2 Which of the following substances have the same shape?

A H_2Se and SF_3

B H_2Se, SF_3 and NO_2

C H_2Se and NO_2

D NO_2 and AsH_3

3 In krypton tetrafluoride (KrF_4) the central atom is surrounded by:

A 2 single bonds, 2 double bonds and 1 lone pair of electrons.

B 4 single bonds, no double bonds and 2 lone pairs of electrons.

C 4 single bonds, no double bonds and no lone pairs of electrons.

D 2 single bonds, two double bonds and 1 lone pair of electrons.

4 What is the name of the intermolecular force in which adjacent molecules are attracted to each other by weak, non-directional temporary dipoles?

____

5 In the group 8 elements (He, Ne, Ar, Kr, Xe and Rn) the trend in vapour pressure increases down the group. What physical property is responsible for this?

____

6 The boiling points of boron trifluoride (BF_3) and methanal (CH_2O) are –100°C and –19°C, respectively. Given that both molecules have the same molecular geometry (trigonal planar), account for this significant difference.

7 Table 13.5.1 provides information on two substances: sulfur dioxide and carbon dioxide.

TABLE 13.5.1 Some physical properties about sulfur dioxide and carbon dioxide

SUBSTANCE	BOILING POINT (°C)	SOLUBILITY IN WATER (gL^{-1})	VAPOUR PRESSURE (MPa)
Sulfur dioxide	−10	94	0.273
Carbon dioxide	−57	1.4	5.73

Use your knowledge of intermolecular forces to explain the information in Table 13.5.1.

9780170412391

8 The graph below shows how boiling point changes going down a group in the periodic table. Refer to Table 13.5.2 to answer the questions. The boiling points for the group 4 hydrides have been plotted on following graph.

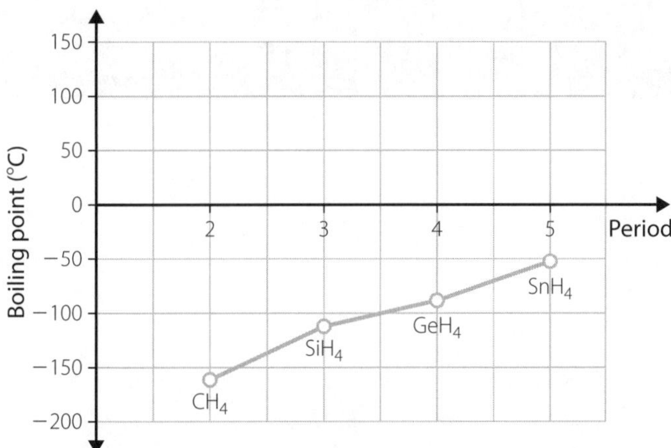

TABLE 13.5.2 Boiling points of some hydrides formed from elements in periods 2–5

PERIOD	COMPOUND	BOILING POINT (°C)	COMPOUND	BOILING POINT (°C)	COMPOUND	BOILING POINT (°C)
2	CH_4	−161	H_2O	100	HF	19.5
3	SiH_4	−112	H_2S	−60	HCl	−85.5
4	GeH_4	−88	H_2Se	−41	HBr	−66
5	SnH_4	−52	H_2Te	−2	HI	−34

a Plot the data for the hydrides of the group 6 and 7 elements on the graph.

b Explain, with reference to intermolecular forces, the trend for *all* the graphs and their relative positions to one another (including the group 4 hydrides, which are already plotted).

____

____

____

____

c For elements in groups 6 and 7, estimate, by extrapolating the appropriate graphs, the boiling point of the hydrides of the elements in period 2. Show your working on the graph.

____

____

____

____

14 Chromatography techniques

Summary

- Chromatography is a separation technique based upon the differential distribution of substances between a stationary phase and a mobile phase.
- The differential distribution between mobile and stationary phases depends on several factors including molecular weight, molecular geometry and intermolecular forces.
- The three main chromatography techniques are: thin layer chromatography (TLC), gas chromatography (GC) and high-performance liquid chromatography (HPLC).
- In TLC, separation of the components in a mixture is achieved by the components spending differing amounts of time dissolved in the solvent (mobile phase) and adsorbed onto the solid surface (stationary phase) of the plate.
- The rate at which a component moves up a plate compared with the rate at which the solvent front moves is called the 'retardation factor' (R_f) and is given by the equation:

$$R_f = \frac{\text{distance travelled by component}}{\text{distance travelled by solvent front}}$$

- GC is a method of separating compounds that can be vaporised without decomposition.
- In GC, the separation of components is achieved by differential distribution between the mobile phase (an inert gas such as helium or an unreactive gas such as nitrogen) and the stationary phase (such as a microscopically thin layer of a liquid polymer on silica beads).
- In GC, the amount of time a substance takes to leave the column and be detected is called the retention time (R_t) and can be used to identify the substance.
- GC can be used as a quantitative technique by measuring the area under the peaks produced by components on a chromatogram.
- HPLC is similar to GC but is better suited to larger organic molecules that are not able to withstand the high temperatures required for vaporisation. HPLC relies on pumps to push the liquid mobile phase through the column.
- As with GC, compounds can be identified using the retention time (R_t) of substances, and quantitative analysis can be achieved with the use of a calibration curve.

9780170412391

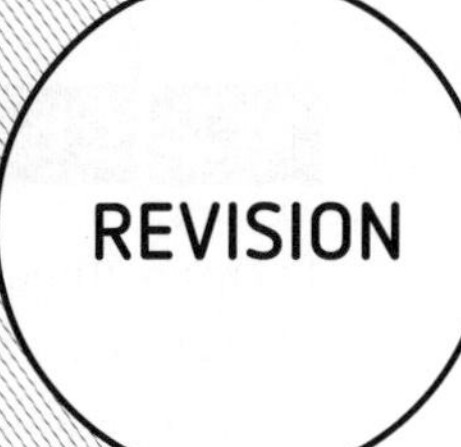

14.1 Important terms

CONCEPT MAP

An incomplete concept map for chromatography techniques is provided with some of the concepts filled in.

1 Use the following concept list to complete the concept map.

Concept list

HPLC	intermolecular forces	qualitative	quantitative
distance moved by component	distance moved by solvent front	GC	

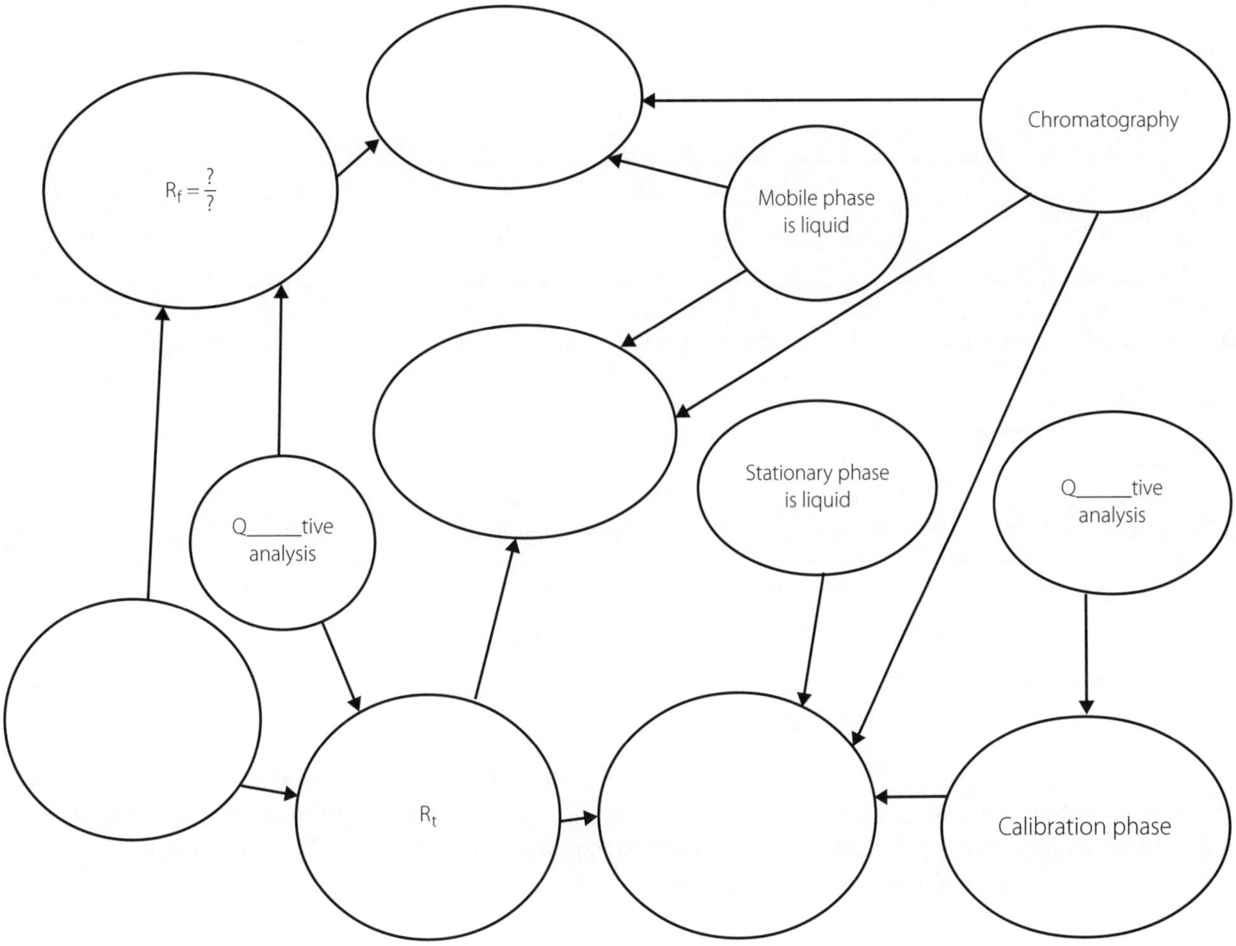

Chromatography

Mobile phase is liquid

$R_f = \frac{?}{?}$

Stationary phase is liquid

Q______tive analysis

Q______tive analysis

R_t

Calibration phase

14.2 Chromatography techniques

There are three broad categories of analytical chromatography: thin layer chromatography (TLC); gas chromatography (GC) (which is sometimes referred to as gas–liquid chromatography (GLC)); and high-performance liquid chromatography (HPLC). Each method relies on the different rates that the analytes in the mobile phase move across the stationary phase. Between them, the three methods can detect a very broad range of substances and each technique has advantages for specific applications.

1 Define the following terms.

a Stationary phase

____

b Retardation factor (R_f)

____

c Desorption

____

d Retention time (R_t)

____

e Calibration curve

____

f Adsorption

____

2 Use the following word list to fill in the blanks in the following sentences.

Word list

analyte	mobile	vapourised	adsorb
pumps	HPLC	intermolecular forces	component
stationary	mobile	gas chromatography	

a In TLC, the ______________ phase is usually a fine powder of silica attached to glass or plastic while the ______________ phase is a solvent in which the ______________ is dissolved.

b Separation of substances depends on the ____________________ present in ________ and how strongly they ______________ onto the surface of the plate.

c In __________________ the ____________ phase is an inert gas. This method is better suited to substances that can be easily __________________ unlike ________________ which is more appropriate for molecules that are unstable at high temperatures. This method requires the use of ____________ to move the ________________ phase through the column.

14.3 Separating components

Different forms of chromatography all rely on the same basic principle: components are separated due to the varying amounts of time they spend in the mobile phase. The varying amounts of time spent in the mobile phase depends largely on the types of intermolecular forces within the components.

Chromatography can be thought of a kind of tug-of-war between the attraction of a component and the stationary and mobile phases.

1 The following sets of substances undergo thin layer chromatography. One set consists of four alkanes dissolved in hexane, another alkane. Alkanes are non-polar substances. The other set consists of group 5 hydrides dissolved in ethanol, a polar solvent.

Using arrows, match the substances in the following diagram with the appropriate R_f values:

Solvent = hexane (C_6H_{14})

C_5H_{12}
C_7H_{16}
$C_{12}H_{26}$
$C_{14}H_{30}$

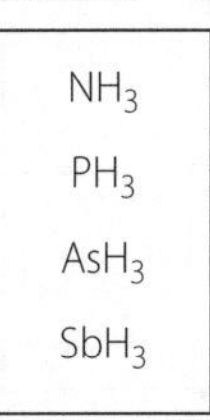

Solvent = ethanol (C_2H_5OH)

NH_3
PH_3
AsH_3
SbH_3

14.4 Inquiry skills

The three types of chromatography are excellent qualitative analytical tools with their use of R_f and R_t values. Follow worked examples 1 and 2 to help solve the described scenarios.

WORKED EXAMPLES

HPLC and GC are invaluable quantitative analytical tools. The output from these techniques, the chromatogram, is generally a series of peaks. The height or area under these peaks can be related directly to the quantities of the components being analysed.

1 A patient is admitted to hospital after having accidentally swallowed a large quantity of turpentine (a common household solvent). A sample of blood was analysed for pinene in a gas chromatograph. Pinene ($C_{10}H_{16}$) is the major component of turpentine. The results are given in Table 14.4.1.

TABLE 14.4.1 Gas chromatography results of the concentration of pinene in blood sample

PINENE CONCENTRATION (ppm)	PEAK AREA (cm^2)
400	1.2
800	2.5
1200	4.1
1600	5.4
2000	6.9
2400	8.3
Sample from blood	5.8

Given that the toxicity of turpentine given as pinene is 1900 ppm, what is the prognosis for the patient?

ANSWER

a Construct a calibration curve for pinene on the grid using the information from Table 14.4.1.

A calibration curve is used to compare the amount of analyte (in this case, pinene) in an unknown sample with known, standard concentrations of sample. In this case the pinene concentration is plotted on the x-axis with peak area on the y-axis. Draw a line of best fit.

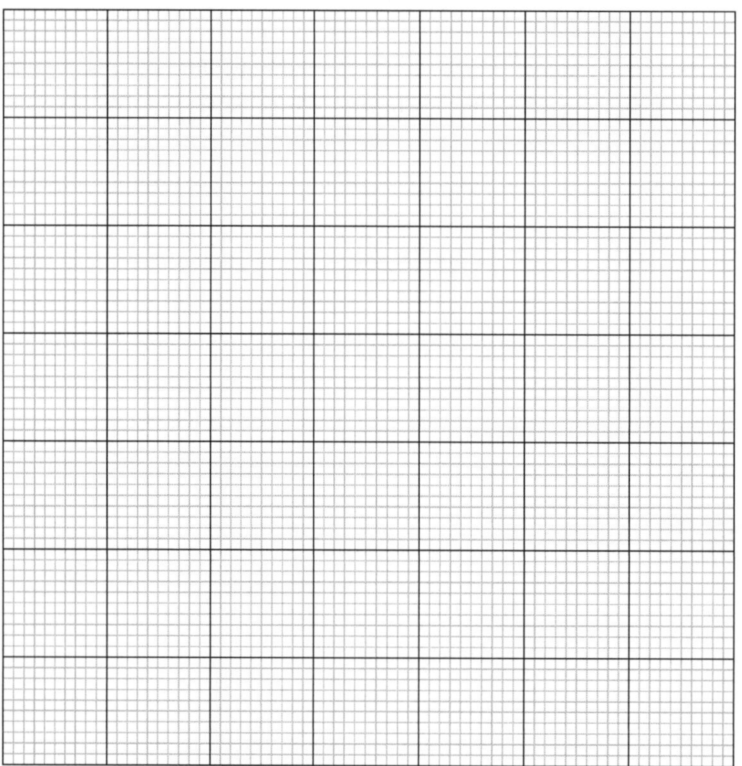

b Determine the concentration of pinene in the blood sample.

On the graph, draw a horizontal line from the point at 5.8 on the y-axis to the line just plotted. From this point draw a vertical line down to the x-axis. This value is the concentration of pinene in the blood sample.

If the graph has been drawn correctly, the value for pinene in the blood sample should be about 1750 ppm.

c Compare the pinene concentration in the blood with the toxicity and comment on the likely prognosis for the patient.

The concentration of pinene in the blood is about 1750 ppm, which is less than the toxic amount, and so the prognosis for the patient is good.

9780170412391

2 A pharmacist had four identical bottles, without labels, containing four different painkillers: aspirin, ibuprofen, paracetamol and tramadol. To identify the painkillers, the pharmacist dissolved a sample of each in an appropriate solvent and applied the solutions to a TLC plate as shown in Figure 14.4.1. Identify each solution by determining the R_f value for each and comparing them with those in the table. Which of the four painkillers is solution 1, 2, 3 and 4?

FIGURE 14.4.1 Thin layer chromatogram of four unknown painkillers with a table of their respective R_f values

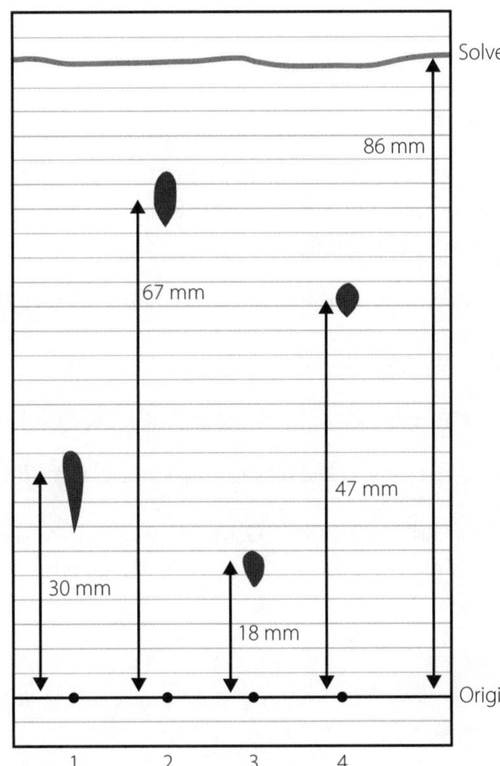

Painkiller	R_f value
Tramadol	0.57
Paracetamol	0.33
Aspirin	0.77
Ibuprofen	0.21

ANSWER

a Measure the distance from the origin to the solvent front.

Distance = 86 mm

b Measure the distance from the origin to the centre of the spot produced by solution 1.

Distance = 30 mm.

c Calculate the R_f value for solution 1.

$$R_f = \frac{\text{distance travelled by component}}{\text{distance travelled by solvent front}} = \frac{30}{86} = 0.35$$

d Compare calculated R_f value with the value given in the table in Figure 14.4.1.

A value of 0.35 corresponds most closely to paracetamol, which has a value of 0.33.

e Repeat steps **a–d** for solutions 2, 3 and 4.

Solution 2: $R_f = \frac{60}{86} = 0.77 = \text{aspirin.}$

Solution 3: $R_f = \frac{21}{86} = 0.21 = \text{ibuprofen.}$

Solution 4: $R_f = \frac{46}{86} = 0.55 = \text{tramadol.}$

EVALUATION

For questions **1** and **2**, circle the best option.

1 A sample of ink was dissolved in an appropriate solvent and spotted onto a TLC plate. After a few hours it was observed that two spots had appeared. Spot 1 was 3.4 cm away from the origin, and spot 2 was 4.2 cm away from the origin. From these results it can be concluded that:

A component 1 was adsorbed more strongly to the plate and so had a smaller R_f than component 2.

B component 1 was adsorbed less strongly to the plate and had a smaller R_f than component 2.

C component 1 was adsorbed less strongly to the plate and so had a larger R_f than component 2.

D component 1 was adsorbed more strongly to the plate and so had a larger R_f than component 2.

2 If a sample of carbon monoxide (CO), carbon dioxide (CO_2), methane (CH_4) and ethane (C_2H_6) was going to be analysed using gas chromatography, choose the most appropriate carrier gas.

A Hydrogen

B Nitrogen

C Oxygen

D Water vapour

3 What is the name of the process by which a component being separated in a HPLC column moves out of the stationary phase and into the mobile phase?

____

4 If a mixture containing various types of alcohol (each one containing the –OH group) is to be analysed using TLC, what kind of solvent should be used, polar or non-polar?

____

5 Using your knowledge of solubility, explain your answer to question **4**.

____

____

____

____

____

9780170412391

6 When determining the quantity of a component using gas chromatography, why must a calibration curve be produced?

7 In the determination of organic compounds with low thermal stabilities, why would HPLC be the best method of choice over TLC and GC?

8 Some black ink was dissolved in ethanol, spotted onto a TLC plate and left to develop. The resulting plate showed that the solvent front had moved 7.4 cm and different coloured spots had appeared at different distances along the plate. The R_f values for the different coloured spots were 0.14, 0.39, 0.42 and 0.78.

Use the information given to fill in the blank plate and mark the position of the four different-coloured spots.

Origin

15 Gases

LEARNING

Summary

- A gas is a substance that exists in a state that has no volume or shape.
- A gas takes up the volume of its container but can be compressed into a smaller volume because the actual volume of gas particles is a tiny fraction of the total volume of the container.
- Avogadro's hypothesis states that equal volumes of gas contain the same number of particles.
- At standard temperature and pressure (STP – 0°C, 1 atmosphere (atm)) one mole of any gas occupies a volume of 22.4 L.
- At standard laboratory conditions (SLC – 25°C, 1 atm) one mole of any gas occupies a volume of 24.5 L.
- The physical properties of gases can be explained by the kinetic theory of gases.
- Gases that conform to the kinetic theory are called 'ideal gases'.
- The SI unit of pressure is the pascal (Pa), and is equivalent to a force of $1\,N\,m^2$.
- Pressure can be expressed in various ways; for example:

  $$1\,atm = 101.3\,kPa = 760\,mm\,Hg$$

- In gas-producing chemical reactions, Avogadro's hypothesis can be used to determine the volume of gas produced.
- Intermolecular forces affect the physical properties of covalent substances such as melting and boiling points, solubility and vapour pressure.

9780170412391

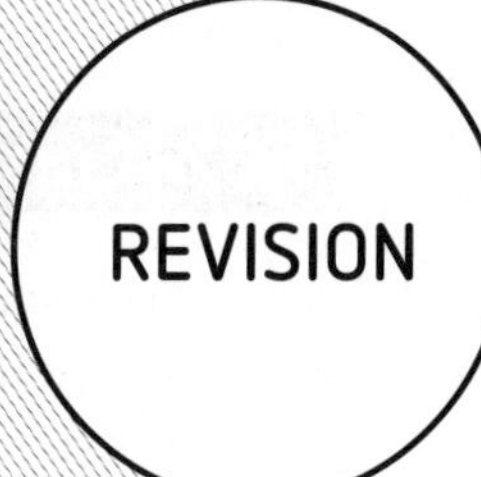

15.1 Important terms

CONCEPT MAP

An incomplete concept map for gases is provided with some of the concepts filled in.

1 Use the terms in the following concept list to complete the concept map.

Concept list

22.4 L	boiling point	Pa	vapour pressure
24.5 L	elastic collisions	Particles are far apart	
As temperature increases, particles move faster	Intermolecular forces are negligible	Particles move in straight lines	
Avogadro's hypothesis	Melting point between particles and the container	Pressure is due to collisions	

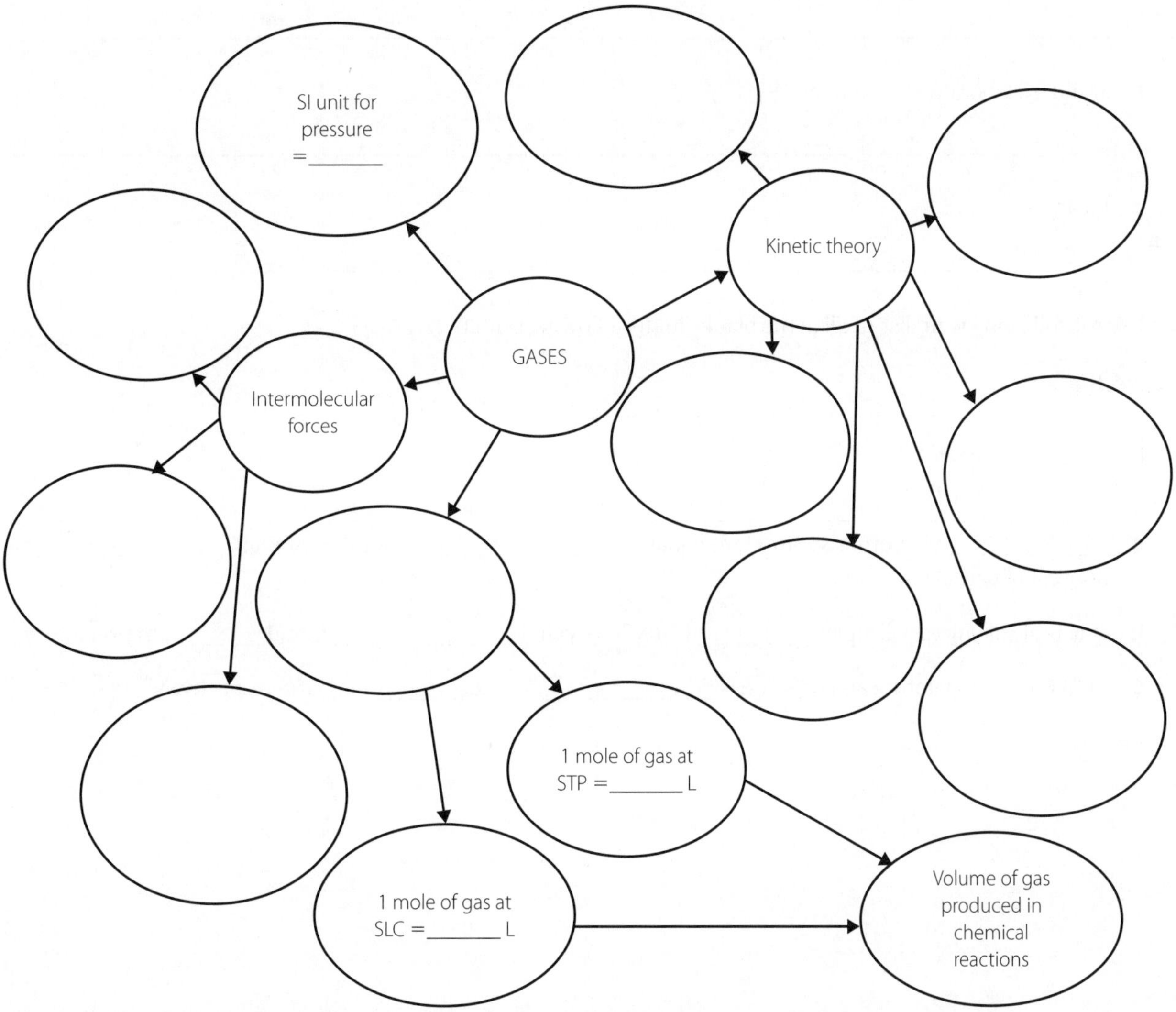

SI unit for pressure =________

Kinetic theory

GASES

Intermolecular forces

1 mole of gas at STP =________ L

1 mole of gas at SLC =________ L

Volume of gas produced in chemical reactions

15.2 Gases

A gas is a substance that has no fixed volume or shape, the 'volume' of a gas merely refers to the volume of the container that holds the gas and the actual volume occupied by the particles of gas is much less than the volume of the container.

Avogadro's hypothesis states that equal volumes of gas contain the same number of particles. Whether the gas is small and light, such as helium, or large and heavy, such as perfluorobutane, is irrelevant because gas particles are minute compared to the distance between them.

1 Define the following terms.

a Vapour pressure

____

b Molar volume

____

c Standard temperature and pressure

____

d Avogadro's hypothesis

____

e Pascal

____

f Standard laboratory conditions

____

g Ideal gas

____

2 Use the following word list to fill in the blanks in the following sentences.

Word list

25	Avogadro	equal	0
1	volumes	22.4	1

a ______________'s hypothesis states that equal ____________________ of gases contain ________________ number of particles.

b One mole of any gas occupies _________ L at STP, which is ______________°C and ______ atm pressure.

c At SLC the temperature and pressure is ______________°C and ___________ atm respectively.

15.3 Kinetic theory

The kinetic theory of gases is used to explain the physical behaviour of gases and defines the behaviour of an ideal gas. While no gas is ideal, under standard conditions, the behaviour of most gases approximates to an ideal gas.

1 Use the following word list to fill in the blanks in the following sentences.

Word list

large	noble gases	negligible
container	energy	elastic

a The kinetic theory of gases states that gases consist of molecules (except for ________________ which are atoms) that move in constant, random straight-line motion.

b The average distance between gas molecules is very __________ compared to the size of the molecule.

c Intermolecular forces are ________________.

d Collisions between molecules are ________________ which means no ______________ is lost during collisions.

e Pressure is due to collisions between gas molecules and the ____________________________.

2 Using arrows on the following diagram, match the values for atmosphere (atm) with the appropriate mm Hg and kPa values. The three columns of numbers represent gas pressures in units of mm Hg, kPa and atm. One set has been completed.

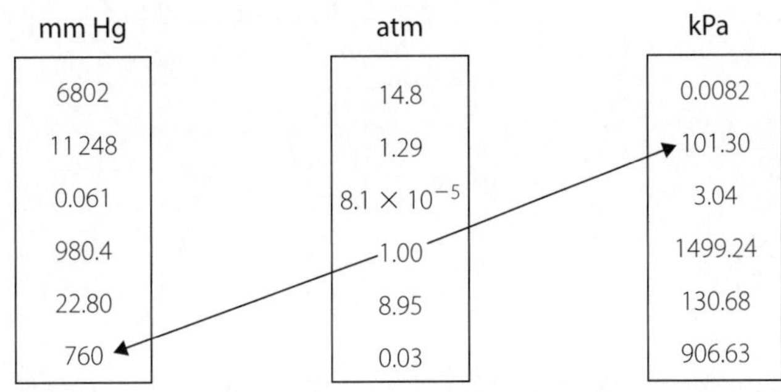

15.4 Inquiry skills: Using mathematical representations

Using the concept of the molar volume of gases, mass/volume relationships in chemical reactions that involve gases to determine. Follow the worked example to calculate the volume of gas produced.

WORKED EXAMPLE

What volume of carbon dioxide gas is produced when 29.65 g of calcium carbonate ($CaCO_3$) reacts with an excess of hydrochloric acid (HCl) under standard laboratory conditions (SLC)?

ANSWER

a Write the balanced equation.

$$CaCO_3(s) + 2HCl(aq) \rightarrow CaCl_2(aq) + CO_2(g) + H_2O(l)$$

b Convert the mass of $CaCO_3$ (29.65 g) to moles.

$$\frac{29.65}{100} = 0.2965 \text{ moles}$$

c Determine the mole ratio from the balanced equation and hence the number of moles of carbon dioxide gas produced.

$$1 \text{ mole of } CaCO_3 : 1 \text{ mole of } CO_2$$

Therefore, the mole ratio is 0.2965 mol $CaCO_3$: 0.2965 mol CO_2.

d Use Avogadro's hypothesis to determine the volume of carbon dioxide gas produced (V).

$$\text{Number of moles of gas at SLC} = \frac{V}{V(1 \text{ mol of gas at SLC})}$$

$$n = \frac{V}{24.5}$$

$$0.2965 = \frac{V}{24.5}$$

Therefore, $V = 0.2965 \times 24.5 = 7.26$ L.

9780170412391

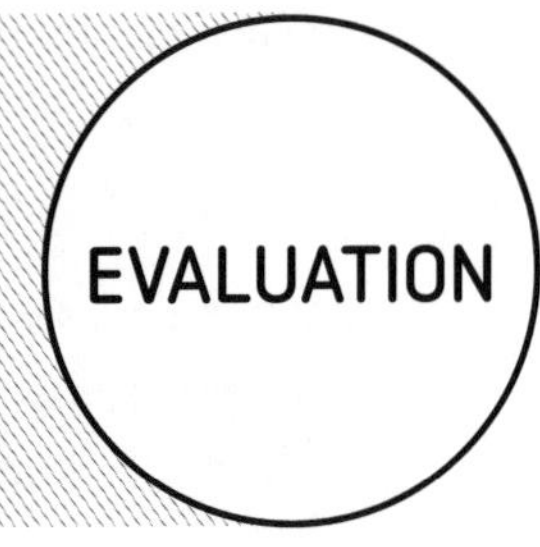

For questions **1** and **2**, circle the best option.

1 If the temperature of an enclosed gas is 423.68 K, what is its temperature in degrees Celsius (°C)?

A 696.68

B 23.68

C 150.68

D 273.68

2 What mass of zinc(II) chloride ($ZnCl_2$) is formed when 13.96 L of chlorine gas reacts with an excess of zinc at 25°C?

A 38.87 g

B 155.48 g

C 84.99 g

D 76.77 g

3 What property of a gas would affect the rate at which it diffuses throughout an area?

____

4 If an inflated balloon is placed into a refrigerator overnight, why does it get smaller?

____

____

5 An experiment is carried out in which various masses of calcium carbonate are reacted with ethanoic acid and the volume of carbon dioxide gas produced is measured according to Figure 15.5.1.

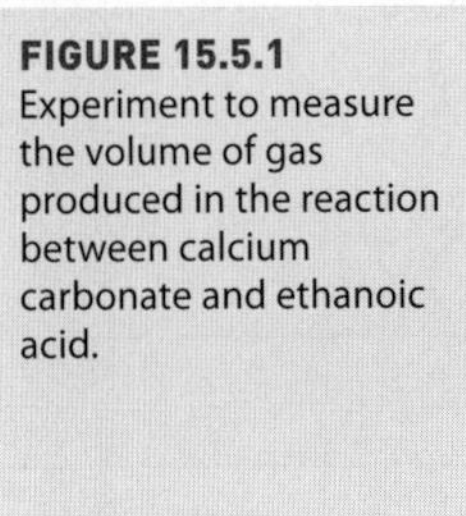

FIGURE 15.5.1 Experiment to measure the volume of gas produced in the reaction between calcium carbonate and ethanoic acid.

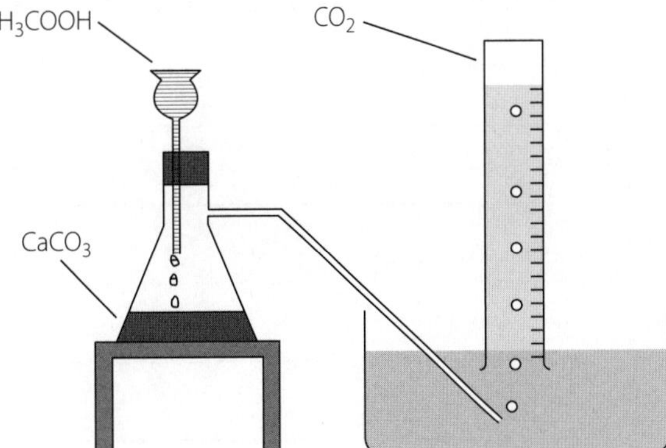

The reaction that takes place is given by the following equation.

$$CaCO_3(s) + 2CH_3COOH(aq) \rightarrow (CH_3COO)_2Ca(aq) + CO_2(g) + H_2O(l)$$

The results are summarised in Table 15.5.1.

TABLE 15.1.1 Volume of gas produced in the reaction between calcium carbonate and ethanoic acid

MASS OF CALCIUM CARBONATE (g)	VOLUME OF CARBON DIOXIDE (mL)
0.10	21.9
0.15	37.2
0.20	48.5
0.25	61.5
0.30	74.0

a Plot a graph of the volume of carbon dioxide against mass of calcium carbonate, drawing a line of best fit passing through the origin.

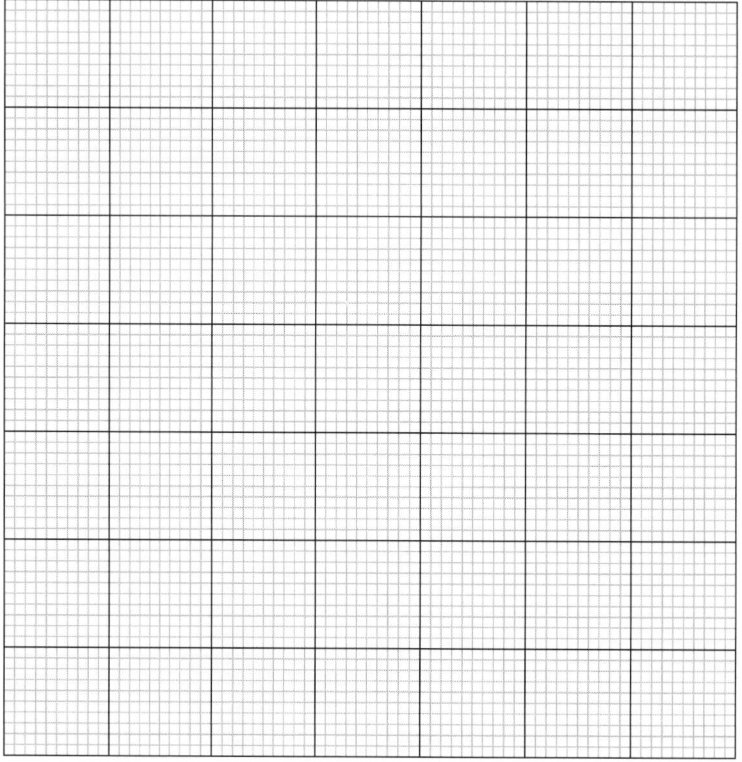

b Use the graph to determine the molar volume of carbon dioxide (and therefore any gas) at SLC, showing your calculations.

9780170412391

c Given that the accepted value for molar volume at SLC is 24.5 L, calculate the percentage error for this value.

d Suggest two possible reasons that might account for this error.

16 Aqueous solutions and molarity

LEARNING

Summary

- Water has unique physical and chemical properties. These include high specific heat capacity, high melting and boiling points, high latent heat of fusion and vaporisation.
- Water's unique properties arise partly from the strength of the intermolecular forces between the water molecules.
- Intermolecular forces between molecules are much weaker than covalent bonds within molecules.
- Hydrogen bonding is responsible for the unique properties of water.
- Water is polar. It dissolves polar and charged substances by forming ion–dipole, dipole–dipole or hydrogen bonds with the substance.
- Non-polar substances dissolve in non-polar solvents but do not dissolve in polar solvents. Polar substances dissolve in polar solvents but do not dissolve in non-polar solvents.
- A solution is a homogenous mixture of one substance (the solute) dissolved in another (the solvent).
- A saturated solution is one in which no more of a particular solute can dissolve in a given quantity of solvent. A solution with less than this amount dissolved in it is an unsaturated solution.
- A supersaturated solution is unstable and has more than the maximum amount of solute dissolved in the quantity of solvent. It has a higher amount of solute than the saturated solution.
- Concentration is a measure of the amount of solute in solution.
- Units for concentration include: g/100 g, % v/v (percentage by volume), % w/w (percentage by weight), ppm (parts per million, grams of solute in one million grams of solution).
- In most chemical contexts, concentration is expressed as M or $mol\,L^{-1}$ (moles per litre, 'molar' or molarity).

16.1 Important terms

1 Complete the crossword using the clues on page 162.

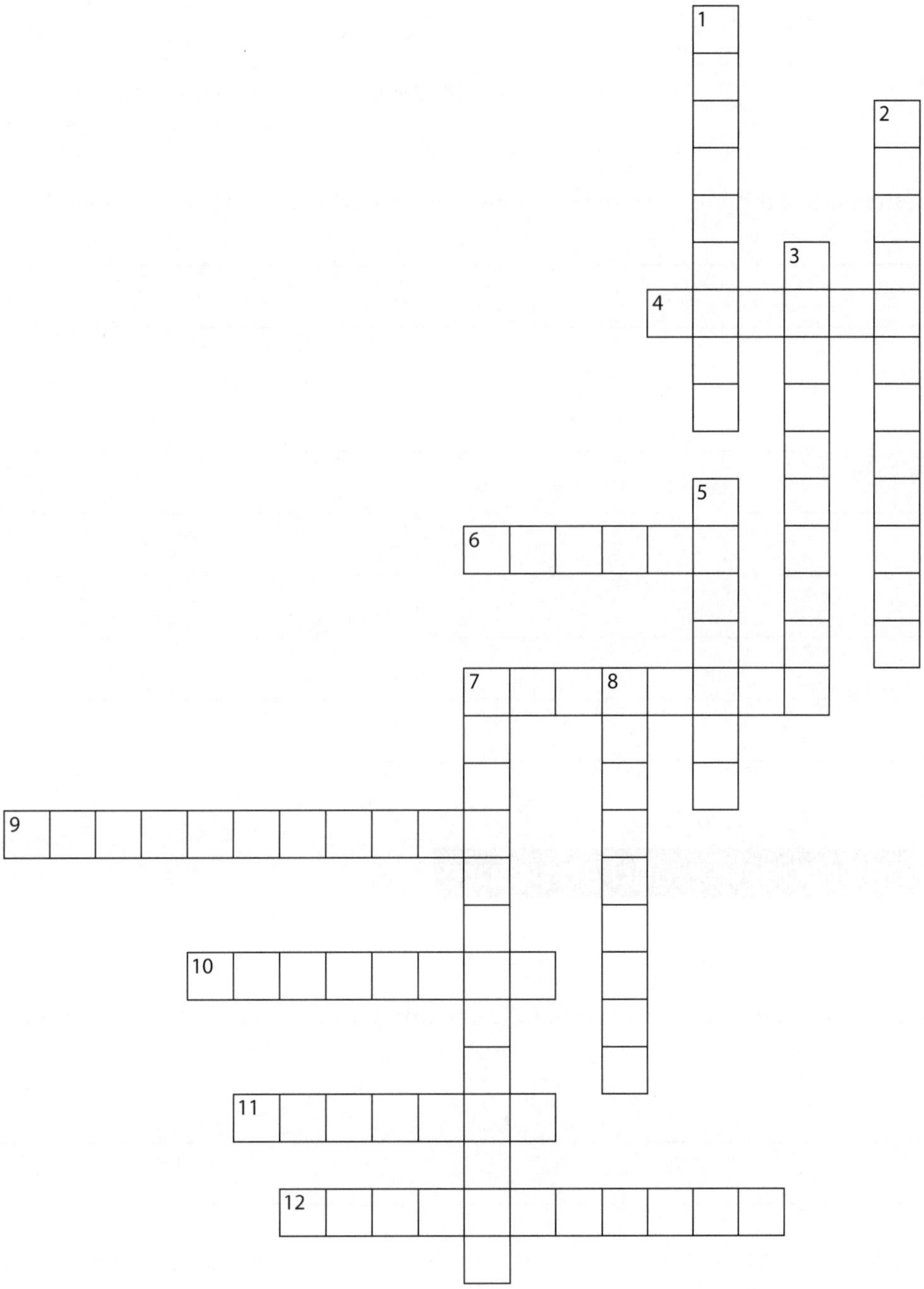

1
2
3
4
5
6
7
8
9
10
11
12

Across	Down
4 A name for the gas phase **6** The substance that dissolves **7** Heat ________ – the energy required to raise the temperature of a substance by 1 degree **9** Water 'loving' **10** Moles per litre **11** The mass per unit volume of a substance **12** Water 'hating'	**1** ____________ action – the ability of a liquid to flow through narrow spaces in opposition to gravity **2** Latent heat of ____________ – the energy required to convert a liquid into a gas **3** The extent to which a solute dissolves in a solvent **5** Surface ________ – the attraction between water molecules on the surface of a liquid **7** Amount of substance per unit volume **8** Ionic solids with no water present

2 Write a short summary of the properties and structure of water using the terms in the crossword.

16.2 Water properties and structure

QUESTIONS

1 Consider the structure of ethane. Do you think it will form hydrogen bonds with water? Explain your answer.

2 Consider the structure of ethanol. Will ethanol form hydrogen bonds with other ethanol molecules, with water or with both? Explain your answer.

3 Explain how hydrogen bonding is responsible for the high boiling point of water.

4 The *intra*molecular forces in water are:

5 The *inter*molecular forces in water are:

6 The bonding that is broken when you turn water into steam is:

16.3 Solutes, solvents, solutions and concentration

1 Identify the solute and solvent in a dilute aqueous solution of sodium chloride.

2 The concentration of a potassium chloride (KCl) solution is given as 7.46 g KCl per 100 g H_2O at 20°C. Determine the concentration of this solution in the following units, stating where your value is only approximate and giving reasons for this.

a Mass of solute per volume of solvent (w/v%).

b Mass percentage (w/w%).

c Moles of solute per litre of solution (i.e. molarity, mol L^{-1}).

3 The recommended minimum level of vitamin C (ascorbic acid, $C_6H_8O_6$) in blood for prevention of scurvy is about 0.3 mg 100 mL^{-1} of blood. State the concentration of vitamin C in blood in the following units.

a Mass percentage (w/w%).

____

____

____

b Moles of solute per litre of solution (i.e. molarity, mol L^{-1}).

____

____

____

9780170412391

For Questions **1** to **4**, circle the best option.

1 Water is referred to as the universal solvent because:

A it is found throughout the universe.

B all known substances dissolve in water.

C many different substances dissolve in water.

D it covers more than half of the Earth's surface.

2 Solubility refers to the:

A size of the particles that make up a substance.

B grams of solute per 100 g of solvent.

C number of moles of solvent dissolved per litre of solution.

D concentration of a solute that is dissolved in a solvent.

3 An unsaturated solution is one that:

A can dissolve more solute at the current conditions.

B will form a precipitate with all of its dissolved solute.

C can dissolve more solute only if heated.

D can dissolve more solute only if the pressure is increased.

4 Identify the most concentrated solution of sodium sulfate (Na_2SO_4).

A 1.0 $mol\,L^{-1}$ Na_2SO_4

B 2.0 $mol\,L^{-1}$ Na_2SO_4

C 3.0 $mol\,L^{-1}$ Na_2SO_4

D 4.0 $mol\,L^{-1}$ Na_2SO_4

5 Saline (NaCl) solutions are used as an eyewash and to help keep wounds germ free. A typical saline solution has a concentration of 0.90% m/v.

a If a capsule containing saline solution has a volume of 5.0 mL, what mass of sodium chloride will it contain?

b Distilled water (495 mL) is added to 5.0 mL of an 8.0 mol L^{-1} solution of sodium chloride. What is the concentration of the solution formed?

9780170412391

17 Identifying ions in solution

LEARNING

Summary

- The solubility of ionic solids in water can vary. Whether an ionic solid will dissolve in water or not can be predicted using the solubility rules.
- An insoluble solid formed when two solutions are mixed is called a 'precipitate'.
- Precipitation reactions can be represented by ionic equations that do not include the spectator ions.
- The identity of specific ions in a solution can be determined using evidence from chemical reactions, including precipitation reactions

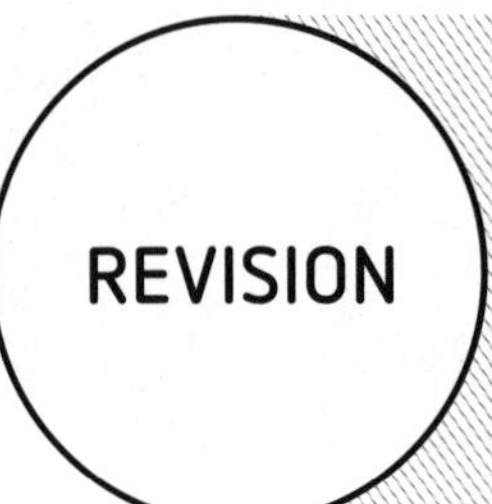

17.1 Important terms

1 Complete the crossword using the clues given.

Across	Down
5 A negatively charged ion	**1** An __________ equation contains all reacting species
6 __________ ion – an ion that remains in solution throughout a reaction	**2** An atom that has gained or lost an electron and become charged
9 A substance that does not dissolve	**3** A positively charged ion
	4 An _________ equation does not contain spectator ions
	7 A solid produced by the reaction between two solutions
	8 A mixture of a liquid and an insoluble solid, where the solid is evenly distributed throughout the liquid

2 Write a short summary about identifying ions in solution using the terms in the crossword.

17.2 Using the solubility rules

TABLE 17.2.1 Some rules for predicting the solubility of ionic compounds

SOLUBLE ANIONS	EXCEPTIONS
NO_3^-	None
CH_3COO^-	Ag^+ slightly soluble
Cl^-	Ag^+ insoluble; Pb^{2+} slightly soluble
Br^-	Ag^+ insoluble; Pb^{2+} slightly soluble
I^-	Ag^+, Pb^{2+} insoluble
SO_4^{2-}	Ba^{2+}, Pb^{2+}, Sr^{2+} insoluble; Ag^+, Ca^{2+} slightly soluble
INSOLUBLE ANIONS	**EXCEPTIONS**
OH^-	Group 1, NH_4^+, Ba^{2+}, Sr^{2+} soluble; Ca^{2+} slightly soluble
O^{2-}	Group 1, NH_4^+, Ba^{2+}, Sr^{2+}, Ca^{2+} soluble
S^{2-}	Groups 1 and 2, NH_4^+ soluble
CO_3^{2-}	Group 1, NH_4^+ soluble
SO_3^{2-}	Group 1, NH_4^+ soluble
PO_4^{3-}	Group 1, NH_4^+ soluble

1 Use the information in Table 17.2.1 to determine whether the following substances (**a–t**) are soluble or insoluble in water.

a Sodium phosphate

b Silver chloride

c Silver nitrate

d Ammonium chloride

e Barium oxide

f Lead oxide

g Magnesium carbonate

h Ammonium sulfate

i Copper hydroxide

j Sodium sulfide

k Calcium phosphate

l Lead nitrate

m Potassium chloride

n Barium sulfate

o Strontium sulfide

p Copper ethanoate

q Calcium carbonate

r Lithium sulfite

s Ammonium phosphate

t Sodium sulfide

17.3 Predicting and writing chemical reactions

1 For each of the following chemical combinations, state whether a reaction occurs or not. If a reaction occurs, write an ionic equation, a balanced chemical reaction using state symbols and name the precipitate that is formed.

a $AgNO_3 + NaCl$

b $NaNO_3 + KCl$

c $CH_3CO_2Ag + KBr$

d $CH_3CO_2K + LiCl$

e $AgNO_2 + NaI$

f $Na_2CO_3 + LiOH$

g $Na_2SO_4 + Pb(NO_3)_2$

h $NaOH + Cu(NO_3)_2$

i $KOH + LiNO_2$

j $Na_2S + Cd(NO_3)_2$

EVALUATION

For questions **1** and **2**, circle the best option.

1 Which of the following pairs of aqueous solutions would result in the formation of a precipitate?

A K_2SO_4 and $NaNO_3$

B NaOH and KNO_3

C Na_2SO_4 and $Ba(OH)_2$

D K_2SO_4 and NaCl

2 Which one of the following contains only substances that are substantially soluble in water?

A $CaCO_3$, C_8H_{18}, HF

B $(NH_4)_2SO_4$, CH_4, NaCl

C CH_3CH_2OH, $NaNO_3$, NH_3

D NaOH, $MgCl_2$, C_5H_{12}

3 When solutions of lithium carbonate and calcium nitrate are mixed, a white precipitate is formed. Write net ionic and full equations to represent this reaction and name the products that are formed.

____

____

____

18 Solubility

LEARNING

Summary

- The temperature of a solution alters the rate at which a substance will dissolve as well as the amount that dissolves.
- A solute dissolves in water when the energy of the bonds it forms with water is lower than the energy of the bonds between water molecules or between the water molecules and ions of the substance being dissolved.
- The solubility of a solute is the maximum amount in grams that can dissolve in 100 g of the solvent at a given temperature.
- Solubility curves are used to display the solubility of various compounds and how their solubility at different temperatures compare.

18.1 Important terms

1 Complete the crossword using the clues given.

Across

1 ______ energy – possessed by moving objects

3 An ionic solution, which therefore conducts electricity

4 An ion that is surrounded by water molecules

6 The weakest type of intermolecular force

7 A solution that looks as though there is only one substance present

8 Two substances that do not easily mix

9 Forces that exist between molecules

10 A reaction that causes the temperature to rise

11 A reaction that causes the temperature to drop

Down

2 Solubility ________ is a graph showing the variation of solubility with temperature

4 A molecule that does not mix easily with water

5 The process where the anion and cation are split when an ionic compound dissolves

9780170412391

2 Write a short summary about solubility and intermolecular factors using the terms in the crossword.

18.2 Forming solutions

1 Match the correct bond type with the appropriate description using an arrow (or list letter pairs that go together). (Letters can be used more than once.)

a Between water molecules	**A** Polar covalent bonds
b Between oxygen and hydrogen in a water molecule	**B** Nonpolar covalent bonds
c Between water and ethanol	**C** Ionic bonds
d Between Na and Cl	**D** Hydrogen bonds
e Between water and ethane	**E** Dispersion forces
f between ethane molecules	

2 Decide which of the following compounds are soluble in water and explain why they are either soluble or insoluble.

a $MgCl_2$

b C_2H_6

c HCl

d CH_4

e $C_{12}H_{22}O_{11}$

f NaOH

3 Write equations to show how the following compounds dissociate in water.

a NH_4NO_3

b LiOH

c H_2SO_4

4 Why do oil and water not mix? Discuss this in terms of what happens at a molecular level.

9780170412391

18.3 Solubility curves

Figure 18.3.1 shows the dependence of the solubility of a solute on temperature. This representation of solubility is termed a 'solubility curve'.

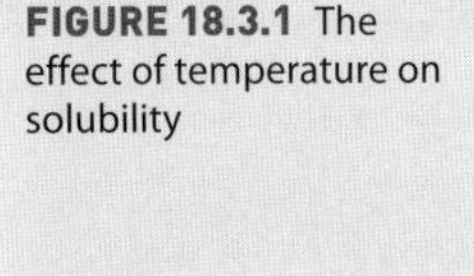

FIGURE 18.3.1 The effect of temperature on solubility

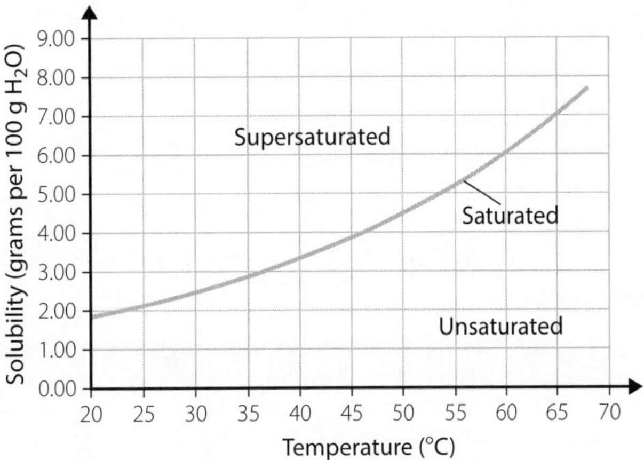

Solubility curves are used to display the solubility of various compounds and how their solubilities at different temperatures compare. Figure 18.3.2 provides an example of the solubilities of several compounds in units of grams of dissolved solute per 100 g of water.

FIGURE 18.3.2 The effect of temperature on the solubilities of several compounds

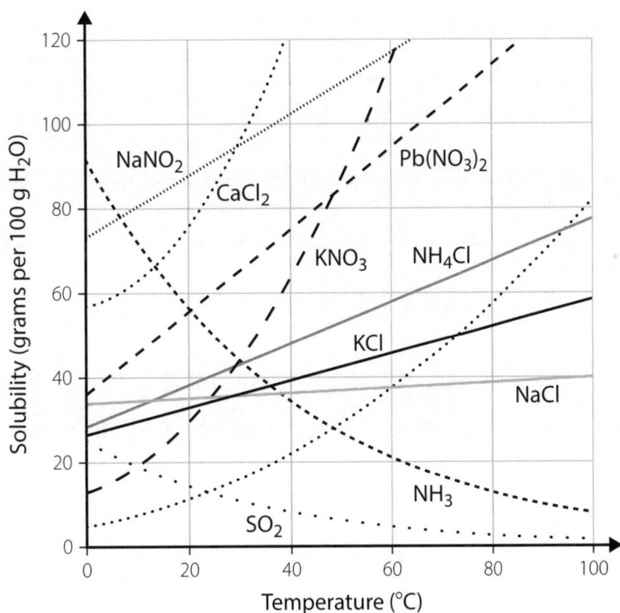

9780170412391

1 Use the solubility curve to determine if the solutions in the following table are saturated or unsaturated. If unsaturated, write how much more solute can be dissolved in the solution. (Note that 100 g of water = 100 mL of water.)

SOLUTION (IN 100 mL OF WATER)	SATURATED OR UNSATURATED?	IF UNSATURATED: HOW MUCH MORE SOLUTE CAN DISSOLVE IN THE SOLUTION (AT THE SAME TEMPERATURE)?
80 g of $NaNO_2$ at 20°C		
40 g of NH_4Cl at 60°C		
60 g of KNO_3 at 40°C		
40 g of NaCl at 80°C		

2 Use the solubility curves in Figure 18.3.2 to complete the following table of solubilities for various compounds at different temperatures.

SUBSTANCE	T (°C)	SOLUBILITY IN 100 g H_2O
$Pb(NO)_2$	20	
$CaCl_2$	32	
NH_3	10, 30, 80	________, ________, ________
NH_4Cl	50	

3 **a** Based on the solubility curves, do gases become more or less soluble with increased temperature?

____

____

____

b How does this compare with the temperature dependence for the solubility of solid compounds?

____

____

____

c Explain the trends in terms of the factors affecting solubility.

____

____

____

____

____

9780170412391

For Questions **1** to **3**, circle the best option.

1 What happens when a supersaturated solution cools down?

A The solution starts to freeze at room temperature.

B The solution will accept even more solute.

C The solute will rise to the top of the solution.

D The solute will precipitate out of the solution.

2 You can make a solute dissolve more quickly in a solvent by:

A adding more solute.

B adding ice.

C heating the solvent.

D removing some solvent.

3 The solubility of calcium chloride ($CaCl_2$) is 77 g per 100 g of H_2O at 20°C. A solution of 77 g of calcium chloride in 100 g of water at 20°C is:

A saturated.

B unsaturated.

C dilute.

D supersaturated.

4 The solubility curve of potassium dichromate ($K_2Cr_2O_7$) is given in Figure 18.4.1.

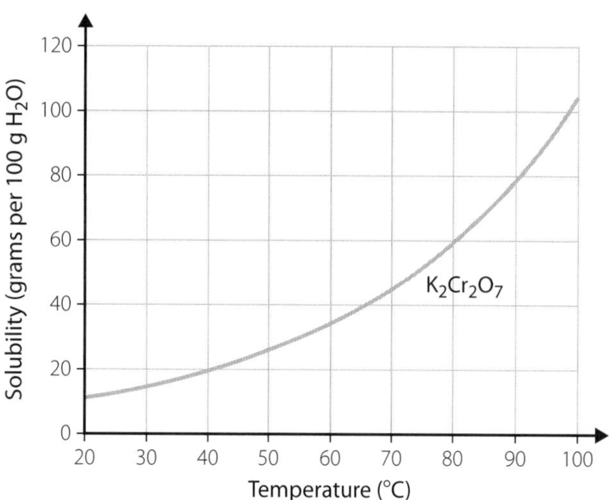

FIGURE 18.4.1 Solubility curve of potassium dichromate

a Use the solubility curve in Figure 18.4.1 to complete the table.

SOLUBILITY OF $K_2Cr_2O_7$	T (°C)	SOLUBILITY
Solubility (g per 100 g of H_2O)	40	
Solubility (mol L^{-1})	40	
Solubility (g per 100 g of H_2O)	80	
Solubility (mol L^{-1})	80	

b Use the solubility curve for potassium dichromate ($K_2Cr_2O_7$) in Figure 18.4.1 to determine if the following solutions are saturated or unsaturated. If the solution is unsaturated, state how much more solute can be dissolved in the solution.

i 40 g of $K_2Cr_2O_7$ in 100 g of H_2O at 60°C

ii 40 g of $K_2Cr_2O_7$ in 100 g of H_2O at 70°C

iii 79 g of $K_2Cr_2O_7$ in 100 g of H_2O at 90°C

9780170412391

19 pH

LEARNING

Summary

- An acid is a substance that releases hydrogen ($H^+(aq)$) ions in solution.
- A base is a substance that contains or releases hydroxide ($OH^-(aq)$) ions in solution.
- The strength of an acid or a base depends on the degree to which it dissociates in solution.
- An alkali is a base that is soluble in water.
- pH, which stands for 'hydrogen power', is a measure of the degree of acidity or alkalinity of a solution.
- The pH scale is used to compare the strengths of acids and bases. It ranges from 1 (strong acids) to 14 (strong bases).
- The pH scale is a logarithmic scale where each integer differs by a factor of 10 from the next.
- The degree of acidity or alkalinity of a substance can be determined by the use of indicators.
- The Arrhenius definition of an acid is a substance that ionises in water to produce H^+ ions.
- The Arrhenius definition of a base is a substance that ionises in water to produce OH^- ions.
- Some strong acids are nitric acid (HNO_3), hydrochloric acid (HCl), and sulfuric acid (H_2SO_4).
- Strong bases include the hydroxides and oxides of groups 1 and 2.

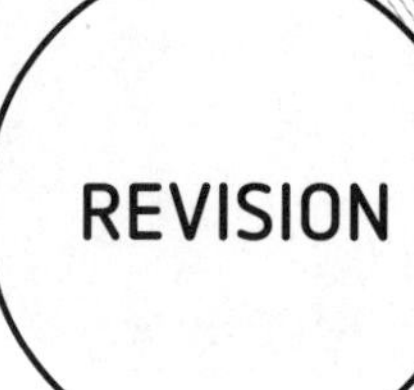

19.1 Important terms

1 Use the terms in the following concept list to complete the concept map.

Concept list

7	CH_3COOH	HCl	NH_3
blue	colourless	NaOH	pink
produce H^+	produce OH^-	red	

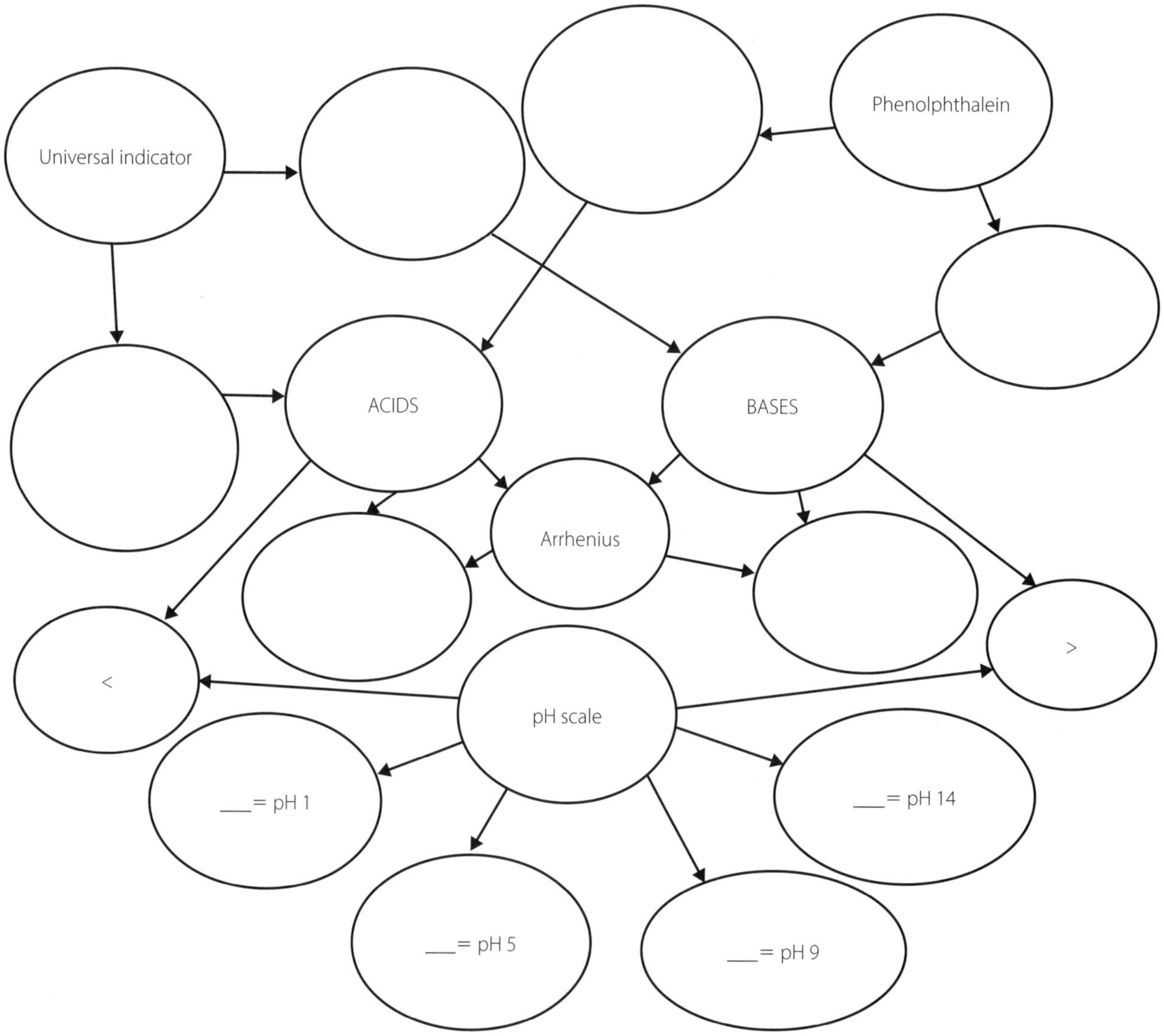

Universal indicator

Phenolphthalein

ACIDS

BASES

Arrhenius

<

>

pH scale

___= pH 1

___= pH 14

___= pH 5

___= pH 9

9780170412391

19.2 pH

The term pH stands for 'hydrogen power' and is used to describe the acidity or alkalinity of substances. Some acids are strong, meaning that they dissociate (or ionise) completely in solution, thereby releasing many H^+ ions into solution. These strong acids include nitric acid (HNO_3), hydrochloric acid (HCl) and sulfuric acid (H_2SO_4). Weak acids dissociate only partially and so release far fewer H^+ ions. Common weak acids include ethanoic acid (acetic acid, CH_3COOH), carbonic acid (H_2CO_3) and phosphoric acid (H_3PO_4).

Strong bases are those that dissociate completely in solution, thereby releasing many OH^- ions into solution. Strong bases include the oxides and hydroxides of groups 1 and 2 elements. Bases that are soluble in water are called alkalis. Weak bases dissociate only partially in solution and include ammonia (NH_3) and methanamine (methylamine, CH_3NH_2).

An amphoteric substance is one that can act as an acid or a base depending on the other reactants present.

1 Define the following terms.

a Acid

____

b Base

____

c Amphoteric

____

d Strong acid

____

e Strong base

____

f Weak acid

____

g Weak base

____

2 Use the following word list to fill in the blanks in the following sentences.

Word list

H^+	hydroxides and/or oxides	base
acts as an acid or base	acetic acid	partially

a A strong ____________ is one that dissociates completely to produce OH^- ions in solution. Strong bases include ________________.

b A weak acid such as ________________ only ________________ dissociates to produce __________ in solution.

c An amphoteric substance is one that ________________________________.

19.3 The pH scale

The pH scale, which commonly ranges from 1 to 14, is used to quantitatively measure the acidity or alkalinity of a substance. It is a logarithmic scale in that each integer has an H^+ ion concentration ten times different from each integer next to it. For example, a substance that has a pH of 4 releases ten times more H^+ ions in solution than a substance with a pH of 5 and 100 times that of a substance with a pH of 6.

Indicators are natural or synthetic substances that can be used to determine the approximate pH of an acid or base by characteristic colour changes.

1 Use the word list provided to fill in the blanks in the following sentences.

Word list

base	less	strong
7	10 000	base

a A ____________ ____________ would have a pH value of 13 whereas a weak ____________ would have a pH of 8.

b A solution resulting from the reaction of the same amounts of acid and base would have a pH of ____________.

c An acid with a pH of 6 would release ____________ times ____________ H^+ ions in solution than a solution with a pH of 2.

19.4 The Arrhenius model

Svante Arrhenius, a Swedish scientist, defined acids as substances that ionise in water to produce H^+ ions and bases as substances that ionise in water to produce OH^- ions.

Acid strength depends on the polarity of the bond containing the H atom – the more polar the bond, the more easily the H atom can break away and be released as an H^+ ion.

Strong bases are those that ionise completely to produce OH^- ions. A substance does not necessarily have to be a hydroxide to produce OH^- ions, but it needs to be able to react with water to produce OH^- ions.

1 Name the following acids and bases and identify them as strong (S) or weak (W). The first one has been completed.

a HNO_3 = nitric acid: S

b CH_3NH_2 = ________________: ___

c CH_3COOH = ________________: ___

d KOH = ________________: ___

e H_2CO_3 = ________________: ___

f NH_3 = ________________: ___

g = potassium hydrogen phthalate: ________________: ___

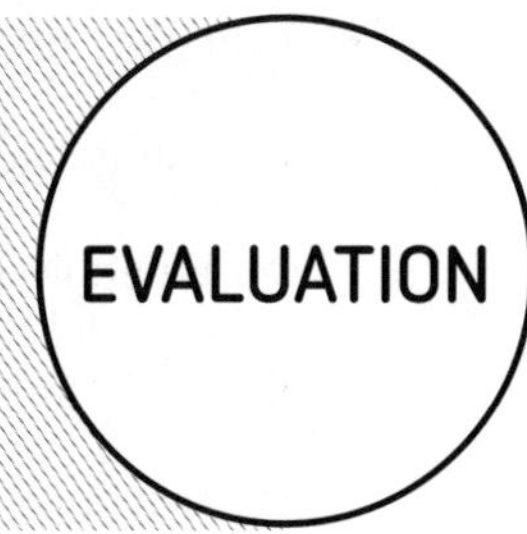

For Questions **1** to **3**, circle the best option.

1 Which of the following best describes vinegar?

A A strong, dilute acid

B A weak, concentrated acid

C A weak, dilute acid

D A strong, concentrated acid

2 Choose the correct explanation for why methanoic (formic) acid is a stronger acid than ethanoic (acetic) acid.

A The O–H bond in ethanoic acid is more polar than in methanoic acid.

B Methanoic acid is more concentrated than ethanoic acid.

C Ethanoic acid loses its H atom more easily than methanoic acid.

D The O–H bond in ethanoic acid is less polar than in methanoic acid.

3 A solution of household ammonia has a pH of 12. This means that it produces:

A 1000 times mores H^+ ions in solution than a substance with a pH of 9.

B 100 times less H^+ ions in solution than a substance with a pH of 14.

C 10 times more H^+ ions in solution than a substance with a pH of 11.

D 10 000 times less H^+ ions in solution than a substance with a pH of 8.

4 Aluminium oxide produces H^+(aq) in the presence of a methanamine (methylamine) solution but produces OH^- (aq) in the presence of ethanoic (acetic) acid. How would aluminium oxide be categorised? Explain your answer.

____

____

____

____

____

5 What is the term given to a substance that changes colour depending on the pH of the solution that it is mixed with?

____

6 Complete the following word equations.

a Sulfuric acid + potassium hydroxide → ______

b Ethanoic acid + ammonia → ______

c Potassium hydrogen phthalate + sodium hydroxide → ______

(See section 19.4, Q1g)

7 The electrical conductivity values of six acids and bases and their reaction with zinc (Zn) metal are given in Table 19.4.1. (Note that the reaction with zinc indicates the presence of an acid.) With reference to Table 19.4.2, suggest the colour change that would occur if a few drops of universal indicator were added to each of the solutions in Table 19.4.1. Explain your decision in each case.

TABLE 19.4.1 The electrical conductivity of six acids and bases and their reaction with zinc metal

SUBSTANCE	REACTION WITH Zn	ELECTRICAL CONDUCTIVITY ($\mu S\,cm^{-1}$)
R	No visible reaction	825
M	Bubbles of colourless gas produced	1 200
L	Bubbles of colourless gas produced	103 000
Q	No visible reaction	53 200
O	Bubbles of colourless gas produced	640
T	No visible reaction	490

a Substance R

______

______

b Substance Q

______

______

c Substance O

______

______

d Substance T

______

______

TABLE 19.4.2 The pH colour changes for universal indicator

pH	COLOUR
1	Red
2	Pink
3	Orange
4	Beige
5	Yellow
6	Light green
7	Green
8	Dark green
9	Turquoise
10	Light blue
11	Blue
12	Dark blue
13	Violet
14	Purple

20 Reactions of acids

LEARNING

Summary

- Acids react with some substances in a predictable manner.
- A neutralisation reaction is one in which just enough base has been added to neutralise an acid to produce a salt and water.
- When an acid reacts with a carbonate, a salt, water and carbon dioxide are produced.
- When an acid reacts with a hydrogen carbonate, a salt, water and carbon dioxide are produced.
- When an acid reacts with a metal, a salt and hydrogen are produced.
- Although the reaction between an acid and ammonia is a neutralisation reaction, it does not produce water. Instead, the hydrogen from the acid attaches to the ammonia, producing the NH_4^+ ion of the ammonium salt product.
- The ammonium ion (NH_4^+) can act as a weak acid.
- Reactions of acids can be represented several ways: the neutral species equation, the complete ionic equation and the net ionic equation.
- The neutral species equation is the overall equation for the reaction. It shows all the reactants and products as neutral substances.
- The complete ionic equation shows all the ions present in solution.
- Reactions of acids can be generalised as:

$$\text{Acid} + \text{base} \rightarrow \text{a salt} + \text{water}$$

$$\text{Acid} + \text{carbonate} \rightarrow \text{a salt} + \text{water} + \text{carbon dioxide}$$

$$\text{Acid} + \text{hydrogen carbonate} \rightarrow \text{a salt} + \text{water} + \text{carbon dioxide}$$

$$\text{Acid} + \text{metal} \rightarrow \text{a salt} + \text{hydrogen}$$

$$\text{Acid} + \text{ammonia} \rightarrow \text{ammonium salt}$$

9780170412391

20.1 Reactions of acids and bases

The reactions of acids and bases follow set patterns which, once learned, enable the products of these reactions to be predicted.

1 Use the word list provided to fill in the blanks in the following sentences.

Word list

carbon dioxide	acetic acid	water	carbon dioxide
ammonia	water	sulfuric acid	magnesium chloride
hydrogen	phosphoric acid	water	

a Hydrochloric acid + magnesium oxide → ______________________
+ ______________________.

b ______________________ + magnesium carbonate → magnesium sulfate
+ ______________________ + ______________________.

c ______________________ + zinc → zinc acetate + ______________________.

d ______________________ + potassium hydrogen carbonate → potassium phosphate
+ ______________________ + ______________________.

e Nitric acid + ______________________ → ammonium nitrate.

2 Complete and balance the following equations.

a $HNO_3(aq) + MgCO_3(s) \rightarrow$

____

____

b $H_3PO_4(aq) + NH_3(aq) \rightarrow$

____

____

c $H_2CO_3(aq) + Zn(s) \rightarrow$

____

____

d $H_2SO_4(aq) + KHCO_3(aq) \rightarrow$

____

____

e $CH_3COOH(aq) + MgO(s) \rightarrow$

____

____

f $NH_4NO_3(aq) + NaOH(aq) \rightarrow$

____

____

20.2 Inquiry skills: Constructing and using representations

The reactions of acids and bases can be represented using three types of equations:

- neutral species equations
- complete ionic equations
- net ionic equations.

WORKED EXAMPLE

1 Write neutral species, complete ionic and net ionic equations for the reaction between dilute sulfuric acid and dilute potassium hydroxide.

ANSWER

a Write the neutral species equation.

This is the overall equation for the reaction and shows all the reactants and products as neutral substances. It is very important to include state symbols in this equation.

$$H_2SO_4(aq) + 2KOH(aq) \rightarrow K_2SO_4(aq) + 2H_2O(l)$$

Sulfuric acid + potassium hydroxide → potassium sulfate + water

b Write the complete ionic equation.

This equation shows all the ions present in solution. This is when correct assignment of state symbols becomes important because when converting the substances in the neutral species equation to ions, it is only the substances with an (aq) state symbol that are converted.

$$2H^+(aq) + SO_4^{2-}(aq) + 2K^+(aq) + 2OH^-(aq) \rightarrow 2K^+(aq) + SO_4^{2-}(aq) + 2H_2O(l)$$

(Note that the water molecule remained the same because it does not have the (aq) state symbol.)

c Write the net ionic equation.

This equation shows only the reacting ions and the products.

$$2H^+(aq) + 2OH^-(aq) \rightarrow 2H_2O(l)$$

(The $SO_4^{2-}(aq)$ and the $K^+(aq)$ ions remained the same on both sides of the equation and could therefore be cancelled out. These ions did not take part in the reaction and are termed 'spectator ions'.)

At this stage, it is very important to note that each substance has a coefficient of 2 in front of it. These must be cancelled out as shown in the following equation.

$$H^+(aq) + OH^-(aq) \rightarrow H_2O(l)$$

QUESTIONS

1 Write neutral species, complete ionic and net ionic equations for the following reactions.

a Dilute nitric acid and sodium carbonate solution.

b Dilute carbonic acid and ammonia solution.

c Dilute ethanoic acid solution and potassium hydrogen carbonate solution.

d Dilute phosphoric acid solution and magnesium.

e Dilute sodium hydroxide solution and ammonium chloride solution.

EVALUATION

For Questions **1** and **2**, circle the best option.

1 Which of the following represents an acid–base reaction?

A $MgCl_2 + PbSO_4 \rightarrow MgSO_4 + PbCl_2$

B $NH_4OH + NaCl \rightarrow NaOH + NH_4Cl$

C $2HNO_3 + MgSO_4 \rightarrow Mg(NO_3)_2 + H_2SO_4$

D $H_2SO_3 + ZnO \rightarrow ZnSO_3 + H_2O$

2 Which of the following is the correct net ionic equation for the reaction between bromic acid ($HBrO_3$) and potassium hydrogen carbonate?

A $HBrO_3(aq) + HCO_3^-(aq) \rightarrow BrO_3^-(aq) + CO_2(g) + H_2O(l)$

B $H^+(aq) + HCO_3^-(aq) \rightarrow CO_2(g) + H_2O(l)$

C $H^+(aq) + 2HCO_3^-(aq) \rightarrow CO_2(g) + H_2O(l)$

D $2H^+(aq) + HCO_3^-(aq) \rightarrow CO_2(g) + H_2O(l)$

3 If dilute sulfuric acid were mixed with potassium hydrogen carbonate and the resulting solution left to evaporate, what would the resulting crystalline solid be called?

____

4 Some strontium hydroxide is mixed with a dilute acid. If the salt produced was strontium phosphate, what was the acid used?

____

5 An experiment is carried out in which dilute sulfuric acid is added from a burette to a dilute strontium hydroxide solution. The conductivity of the mixture is measured at regular intervals. The apparatus is shown in Figure 20.2.1. The results are given in Table 20.3.1.

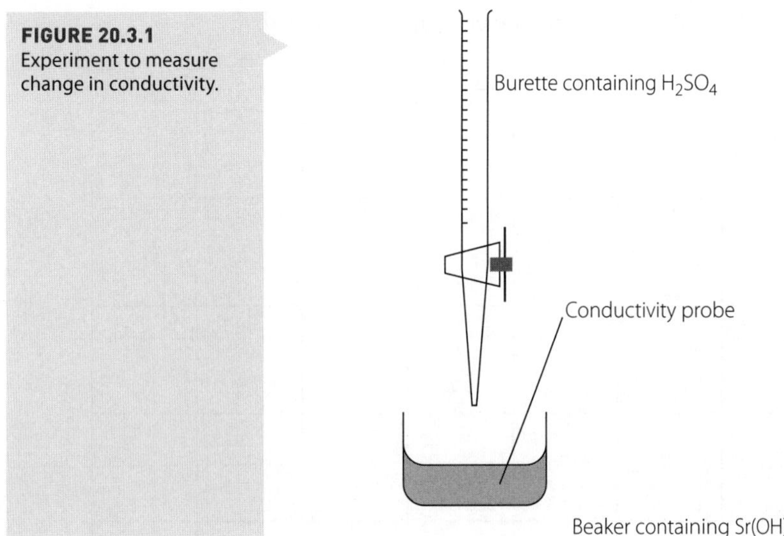

FIGURE 20.3.1 Experiment to measure change in conductivity.

TABLE 20.3.1 Changes in conductivity of a dilute strontium hydroxide ($Sr(OH_2)$) solution after the addition of dilute sulfuric acid (H_2SO_4)

VOLUME OF H_2SO_4 ADDED (mL)	CONDUCTIVITY ($\mu S cm^{-1}$)
0.0	135
1.0	125
2.0	110
3.0	95
4.0	80
5.0	75
6.0	65
7.0	50
8.0	40
9.0	30
10.0	95
11.0	150
12.0	205
13.0	260

The equation for the reaction is:

$$H_2SO_4(aq) + Sr(OH)_2(aq) \rightarrow SrSO_4(s) + 2H_2O(l)$$

a Plot a graph of conductivity against volume of sulfuric acid added.

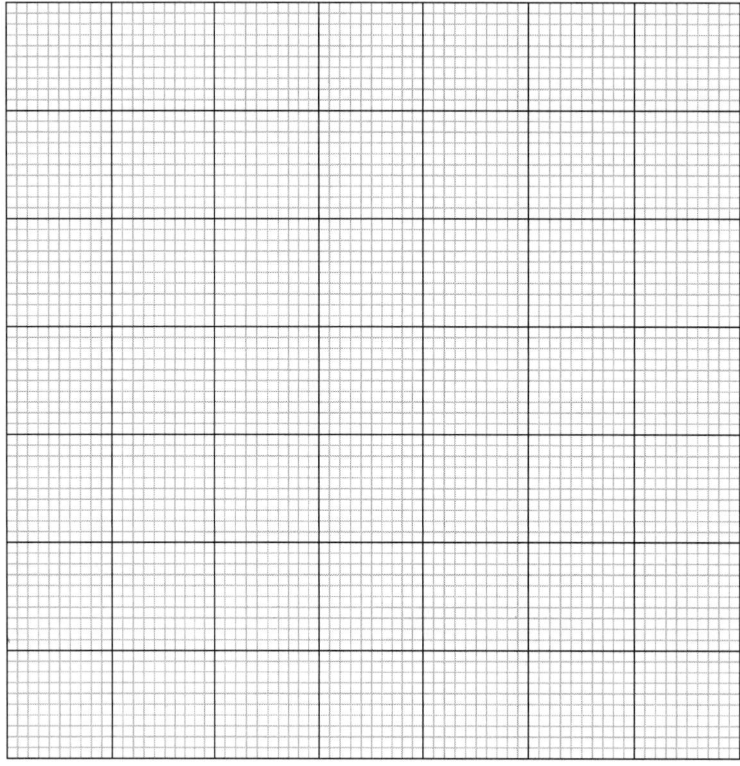

b Determine the endpoint of the reaction (i.e. the volume of sulfuric acid required to neutralise the strontium hydroxide). Show your working on the graph.

c Explain the shape of the graph.

____

____

____

____

____

21 Rates of reactions

LEARNING

Summary

- The rate of a reaction is a measure of how the quantity of either the reactant or the product of the reaction changes with time.
- The rate of a reaction can be determined experimentally in several ways, such as measuring the time taken for a particular amount of reactant to disappear, or product to appear, or alternatively measuring the amount of reactant used or product formed during a specified time interval.
- In a typical graph of the amount of reactant or product against time, the rate is the gradient of the graph.
- For a chemical reaction, three things need to occur.
  - Reactant molecules need to contact one another.
  - Bonds in the reactant molecules need to break.
  - New bonds in the product molecules need to form.
- Collision theory states that for a reaction to occur, the reactants must collide with sufficient energy and in an appropriate orientation. Therefore, any change to the reaction conditions that affects how the reactant molecules collide is likely to affect the rate of the reaction.
- The state of the reaction once the reactant bonds have been broken, but before the product bonds have been formed, is known as the transition state, or activated complex.
- The activation energy is the minimum amount of energy required for a reaction to occur between colliding particles (i.e. to form the activated complex). If the value of the activation energy is high, then the reaction will be slower, as fewer particles will be able to react and vice versa.
- The Maxwell–Boltzmann distribution shows the energies of the particles in a sample at a particular temperature. The particles able to react, whose energy value is above that of the activation energy, can be seen in the following Maxwell–Boltzmann distribution (Figure 21.1).

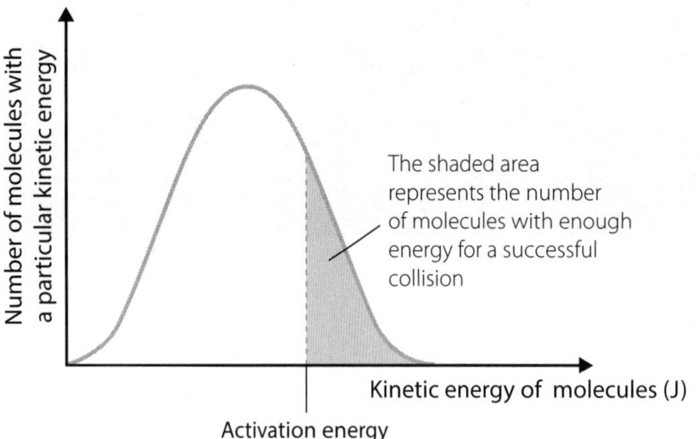

FIGURE 21.1 The Maxwell–Boltzmann distribution shows the transitional kinetic energy of molecules at a particular temperature.

- Energy profile diagrams such as those shown in Figure 21.2 can be used to represent the energy changes occurring during a reaction. The size of the activation energy gives an indication of the likely rate of the reaction.

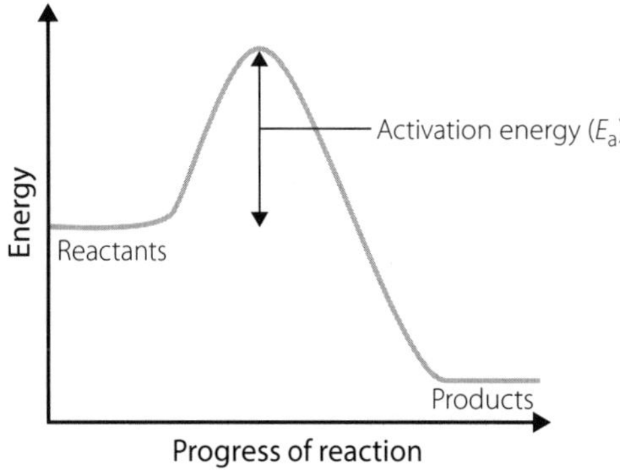

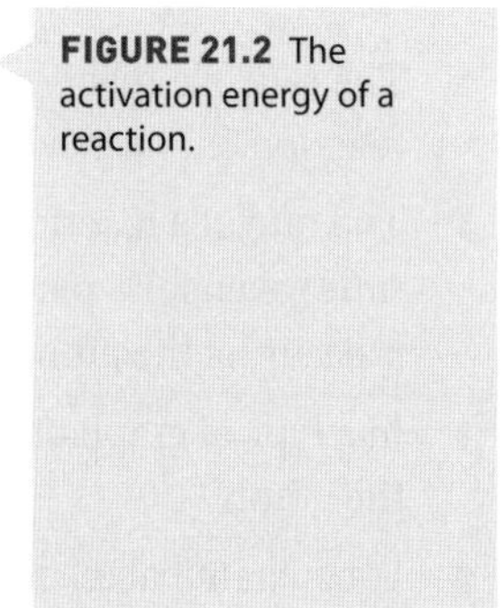

FIGURE 21.2 The activation energy of a reaction.

- Collision theory can be used to explain the effect of temperature, surface area, concentration, and pressure (gaseous systems) on the reaction rate by considering their effect on the rate at which particles are colliding and also on the proportion of these collisions that lead to a reaction.
- Catalysts can alter the rate of a reaction by providing an alternative pathway for the reaction to occur with lower activation energy.
- Enzymes are biological catalysts, complex protein molecules that are critical in regulating and controlling reactions within living organisms. As such they are sensitive to changes in temperature and pH, when their protein structure can become denatured and the enzyme is unable to function.

9780170412391

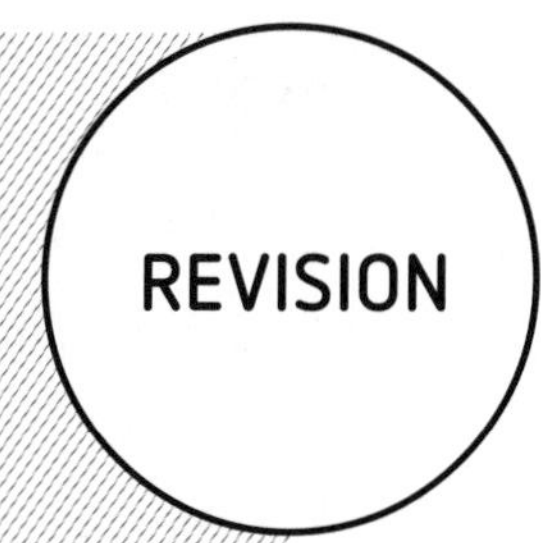

21.1 Important terms

1 Complete the crossword using the clues given on page 198.

Across	Down
3 Substances formed in a chemical reaction	**1** Slope or steepness of a line
4 Type of complex formed when reactant bonds have broken but product bonds have yet to form	**2** Type of energy required to break bonds of reactants
7 Type of theory explaining reaction rates at a molecular level	**5** Amount of solute dissolved in a given volume of solvent
12 Type of graph showing the distribution of energies (or velocities) of particles	**6** Speed of a chemical reaction
13 Total energy possessed by a chemical substance	**8** Type of metal particle with a diameter less than 100 nm that behaves as a whole unit
15 Force exerted by a gas or gases due to collisions of particles with walls of a container	**9** Type of energy possessed by moving objects
16 Type of reaction that absorbs heat from surroundings because reactants have less enthalpy than products	**10** Substance that changes reaction rate without being consumed
18 Type of temporary, unstable state when reactants are forming products	**11** Measure of the average kinetic energy of particles
20 Total area of all exposed surfaces of a solid reactant	**14** Starting substances in a chemical reaction
21 Type of reaction that releases energy to surroundings because products have less energy than reactants	**17** Type of profile diagram showing relative energies during the progress of a reaction
	19 How much one quantity changes with respect to another

2 Write a short summary about atomic structure using the terms in the crossword.

21.2 Constructing energy profile diagrams

The total energy possessed by a chemical substance is called 'enthalpy' or 'heat content', and is given the symbol H. In a reaction, the reactants, activated complex and products each have their own enthalpies. These can be represented in an energy profile diagram.

QUESTIONS

1 Use the information below to construct a labelled energy profile graph of an exothermic reaction.

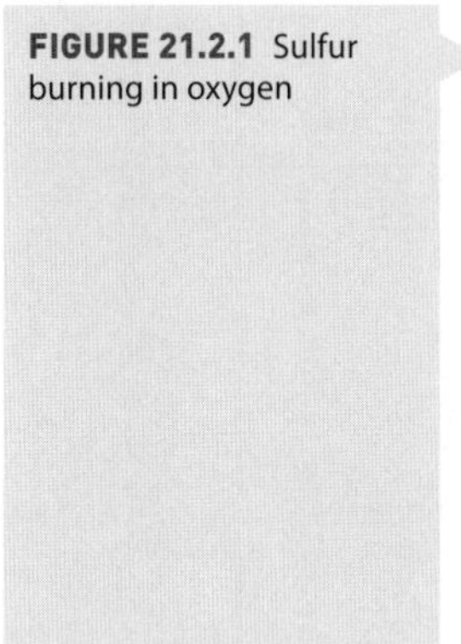

FIGURE 21.2.1 Sulfur burning in oxygen

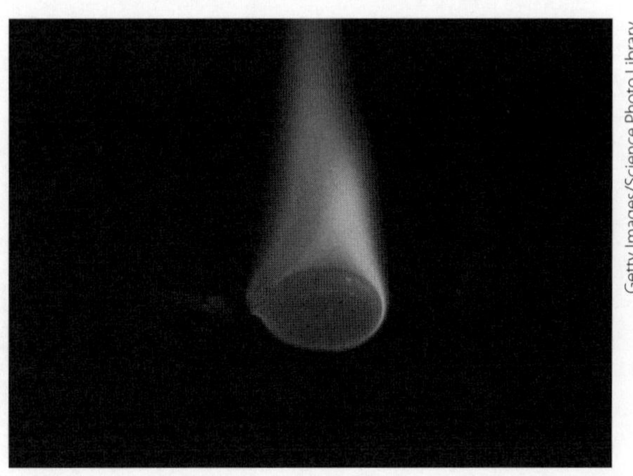

Getty Images/Science Photo Library

When solid sulfur (S) is heated in oxygen gas (O_2), the following reaction occurs:

$$S(s) + O_2(g) \rightarrow SO_2(g)$$

The change in enthalpy when 1 mole of the product is formed is $\Delta H = -2296\,\text{kJ}$.

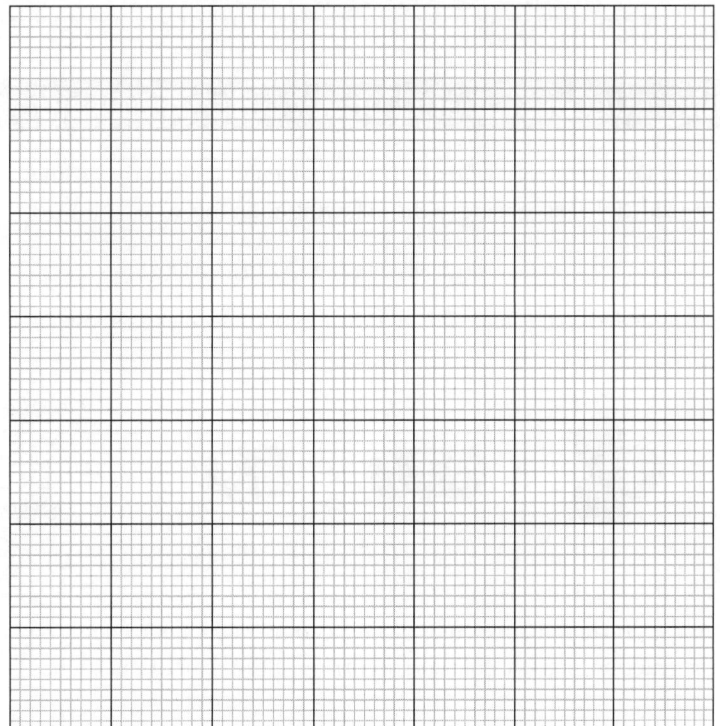

2 Use the information below to construct a labelled energy profile of an endothermic reaction.

When methane is reacted with steam at high temperature and pressure in the presence of a nickel catalyst, carbon monoxide (CO) and hydrogen gas (H_2) are formed according to the following equation.

$$CH_4(g) + H_2O(g) \rightarrow CO(g) + 3H_2(g)$$

The change in enthalpy is $\Delta H = 1206\ \text{kJ mol}^{-1}$.

21.3 Modelling successful and unsuccessful collisions

When carbon monoxide (CO) and nitrogen dioxide (NO_2) molecules collide, they sometime react to form carbon dioxide (CO_2) and nitric oxide (NO) according to the following equation.

$$CO(g) + NO_2(g) \rightarrow CO_2(g) + NO(g)$$

The reaction can be modelled using space-filling drawings of molecules as shown in Figure 21.3.1.

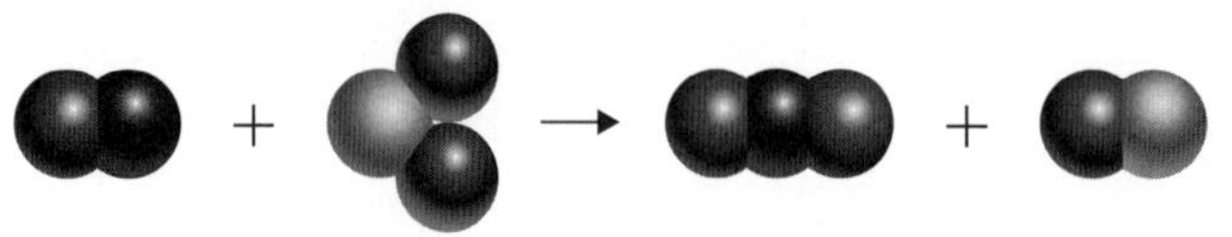

FIGURE 21.3.1 A space-filling drawing of molecules in a chemical reaction. (C = black; N = blue; O = red).

Successful collisions depend on the molecules having a combined kinetic energy equal to or greater than the activation energy for the reaction, and the collision occurring at the right orientation.

9780170412391

1 Using the colour coding in Figure 21.3.1, draw space-filling molecular model diagrams in the following table to illustrate each type of collision (**a–d**). Show the direction and speed of molecules using arrows.

COLLISION TYPE	DIAGRAM	RESULT
a Sufficient energy at the correct orientation		
b Correct orientation but insufficient energy		
c Sufficient energy but wrong orientation		
d Insufficient energy and wrong orientation		

21.4 Effect of altering conditions on reaction rates

1 Draw a sketch to represent each of the actions in the table (**a–h**). Complete the table by describing the effect of the action on the reaction rate and use the collision theory of reactions to explain why the action has a particular effect.

ACTION	EFFECT	EXPLANATION
a Heating reactants		
b Using a powdered form of a solid reactant		
c Increasing the pressure on a gas reactant		
d Cooling a liquid reactant		
e Increasing the concentration of a liquid reactant		
f Adding more water to dissolved reactants		

9780170412391

ACTION	EFFECT	EXPLANATION
g Adding a catalyst to dissolved reactants		
h Increasing the volume of a gas reactant		

21.5 Graphing and interpreting results

In an experiment, 2 g of marble chips (calcium carbonate ($CaCO_3$)) were placed in excess dilute hydrochloric acid (HCl), and the gas produced in the reaction was measured using the apparatus shown in Figure 21.5.1.

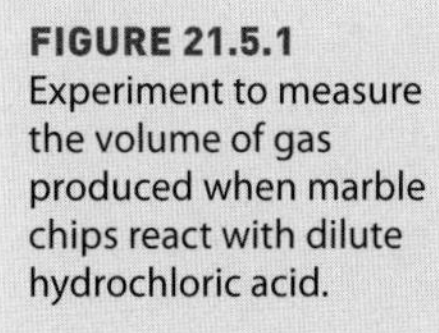

FIGURE 21.5.1 Experiment to measure the volume of gas produced when marble chips react with dilute hydrochloric acid.

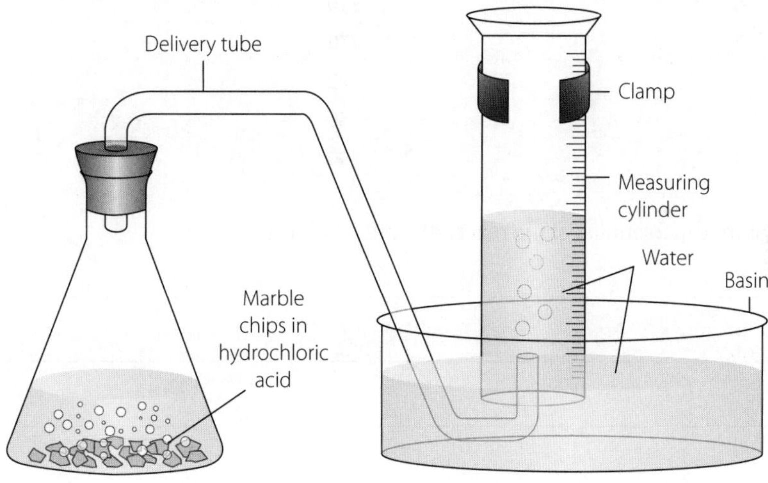

1 Explain why the measuring cylinder was inverted and initially full of water.

2 Describe how you would confirm that the gas produced is carbon dioxide.

The apparatus shown in Figure 21.5.1 was used to measure the gas produced each minute over a 10-minute period, then repeated using 2 g of powdered calcium carbonate and fresh dilute excess hydrochloric acid. The results are shown in Table 21.5.1.

TABLE 21.5.1 Volumes of gas produced in the reaction between dilute hydrochloric acid and the same mass of either marble chips or powdered calcium carbonate

	VOLUME GAS (mL)	
TIME (min)	USING CHIPS	USING POWDER
0	0	0
1	20	70
2	60	130
3	120	200
4	190	280
5	250	370
6	310	430
7	370	460
8	420	470
9	430	470
10	430	470

3 State the formulas of the reactants and products.

4 Write a word equation and a balanced formula equation for the reaction.

9780170412391

5 Plot the reaction progress for the chip and powder on the same graph, using a key. Describe the key.

6 Describe the rate of each reaction.

7 Identify the reaction that went to completion faster.

8 Explain why the reaction in question **7** went faster than the other.

9 Account for why the rate of the faster reaction reached a plateau earlier.

10 Suggest a reason why the amounts of carbon dioxide produced differed.

1 The following figure represents the distribution of energy of a sample of particles taking part in a chemical reaction. The line labelled E_A represents the value of the activation energy for a particular reaction.

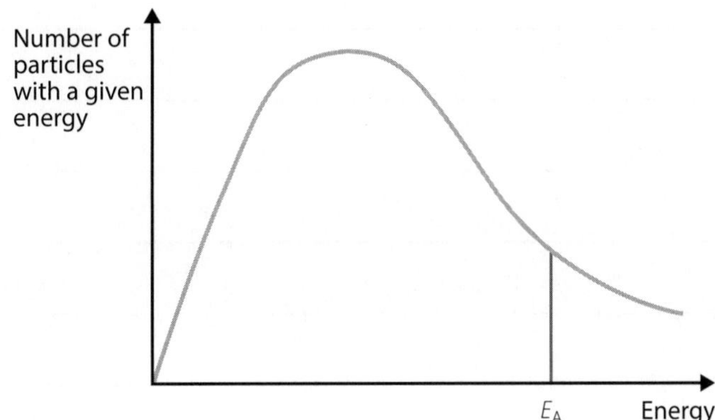

a On the same diagram, draw another line, labelled E_{CAT}, to show what would happen to the activation energy if the reaction were repeated with a catalyst that increased the rate of the reaction.

b Explain, with reference to the figure, how the catalyst causes the rate of the reaction to increase.

2 The figure in Question **1** represents the sample of particles at a particular temperature.

a On the same figure, draw a line, labelled T_2, to indicate how the shape of the graph would look if the temperature were increased.

b Explain, with reference to the figure in question **2**, why increasing the temperature would cause an increase in the rate of a reaction.

3 The rate of a reaction can also be increased by increasing the concentration of the reactant molecules. Explain why this is so and why this change cannot be represented on the figure in Questions **1** and **2**.

9780170412391

CHEMISTRY UNITS 1 AND 2

MULTIPLE CHOICE

For questions **1** to **10**, circle the best option.

Question 1

Which is the smallest subatomic particle in the atom?

A Proton

B Neutron

C Electron

D Nucleus

Question 2

The correct definition for the mass number of an element is the number of:

A protons in the nucleus of an atom of an element.

B protons and neutrons in the nucleus of an atom of an element.

C neutrons in the nucleus of an atom of an element.

D electrons that orbit the nucleus of an atom of an element.

Question 3

An unknown element X has 6 protons and 6 neutrons. Which of the following is an isotope of element X?

A Carbon-12

B Carbon-14

C Boron-11

D Nitrogen

Question 4

Which of the following correctly compares the atomic radius and electronegativity of Cs with Ba?

A Ba has a smaller atomic radius than Cs, and Cs is more electronegative than Ba.

B Cs has a smaller atomic radius and is more electronegative than Ba.

C Ba has a larger atomic radius and is more electronegative than Cs.

D Cs has a larger atomic radius than Ba, and Ba is more electronegative than Cs.

Question 5

What is the closest estimate of the concentration of a mercury sample that produced an absorbance reading of 0.4, as shown in the following figure?

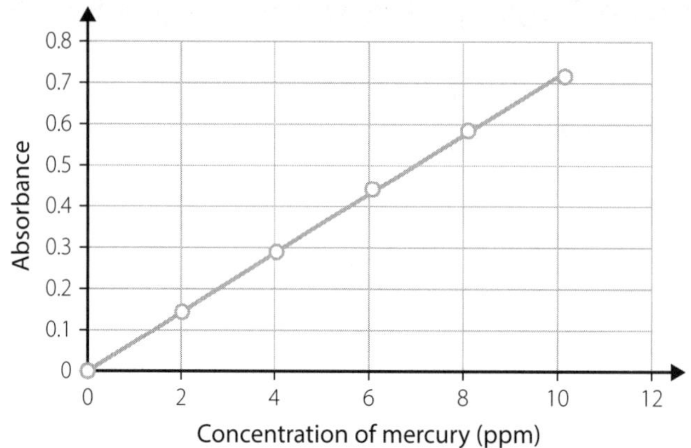

A 0.3 ppm

B 5.0 ppm

C 5.5 ppm

D 6.0 ppm

Question 6

How many moles of oxygen atoms are in 0.98 mol of calcium carbonate?

A 0.98 mol

B 1.96 mol

C 2.94 mol

D 3.92 mol

Question 7

A student added a strip of magnesium with a mass of 0.7 g to a test tube of hydrochloric acid. A rapid stream of gas bubbles was given off and after one minute, all of the magnesium had disappeared. How many moles of magnesium had reacted?

$$Mg + 2HCl \rightarrow MgCl_2 + H_2$$

A 0.015 mol

B 0.03 mol

C 0.7 mol

D 1.4 mol

Question 8

Trigonal planar molecules have:

A four electron pairs; two of which are lone pairs.

B four electron pairs; one of which is a lone pair.

C three electron pairs; one of which is a lone pair.

D three electron pairs, none of which is a lone pair.

9780170412391

Question 9

What is the name of the precipitate formed when KOH and $Mg(NO_3)_2$ are added together? (Hint: refer to the data in the following table of solubilities.)

A Potassium nitrate

B Magnesium hydroxide

C KNO_3

D $Mg(OH)_2$

Soluble anions	Exceptions
NO_3^-	None
CH_3COO^-	Ag^+ slightly soluble
Cl^-	Ag^+ insoluble, Pb^{2+} slightly soluble
Br^-	Ag^+ insoluble, Pb^{2+} slightly soluble
I^-	Ag^+, Pb^{2+} insoluble
SO_4^{2-}	Ba^{2+}, Pb^{2+}, Sr^{2+} insoluble, Ag^+, Ca^{2+} slightly soluble
Insoluble anions	**Exceptions**
OH^-	Group 1, NH_4^+, Ba^{2+}, Sr^{2+} soluble; Ca^{2+} slightly soluble
O^{2-}	Group 1, NH_4^+, Ba^{2+}, Sr^{2+}, Ca^{2+} soluble
S^{2-}	Groups 1 and 2, NH_4^+ soluble
CO_3^{2-}	Group 1, NH_4^+ soluble
SO_3^{2-}	Group 1, NH_4^+ soluble
PO_4^{3-}	Group 1, NH_4^+ soluble

Question 10

Which of the following is closest to the mass of aluminium that must be added to excess HCl to produce 25.7 L of hydrogen gas, at STP?

A 22.4 g

B 31 g

C 62 g

D 21 g

SHORT ANSWER AND COMBINATION-RESPONSE QUESTIONS

Question 1

Explain which reaction (**a** or **b**) is exothermic and which is endothermic.

a $A + B \rightarrow C + D$ $\Delta H = 124\,kJ\,mol^{-1}$

b $E + F \rightarrow G + H$ $\Delta H = -2312\,kJ\,mol^{-1}$

Reaction **a** is: ______________________

Reaction **b** is: ______________________

Question 2

Complete the following table. Calculate the number of neutrons for the three atoms and indicate which of the three atoms are isotopes of each other. The isotopes are: _____ and ___________.

Atom	Atomic number	Mass number	Number of neutrons
A	23	56	
G	22	55	
J	23	55	

Question 3

An unknown solution conducts electricity and turns blue litmus paper red. State two additional properties that the solution might have.

____

____

Question 4

A gas that is confined to a rigid container is heated. Complete the following table to indicate the effect of heating this gas on the following properties.

Property	Increase, decrease or unchanged?
Temperature	
Volume	
Pressure	
Kinetic energy of the gas particles	

Question 5

Draw the electron configuration of chromium, using s, p, d notation.

Question 6

Explain what is meant by the term 'relative molecular mass'.

Question 7

Explain the factors that affect how particles of different sizes and properties are separated in a gas chromatography column.

Question 8

The following figure indicates the boiling points of the hydride compounds of elements in groups 14, 15, 16 and 17 of the periodic table. Provide explanations for the variations in boiling point.

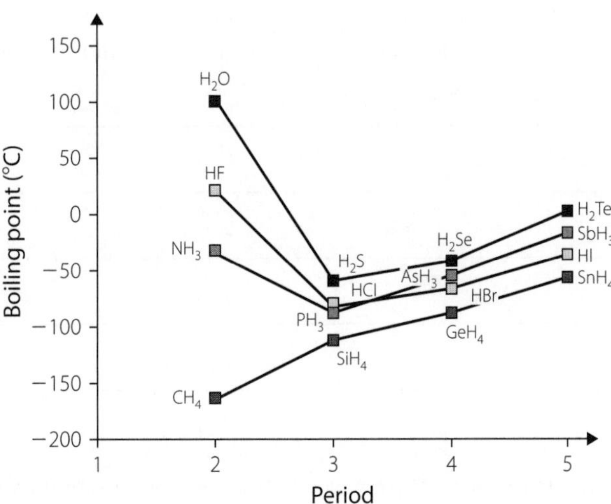

a Which factors affect the boiling points of these compounds?

b Why do the boiling points of the hydrides of the elements in periods 3, 4 and 5 increase with increasing period number?

c Why is the pattern in the hydrides of group 14 elements different to that in the other three groups?

d Boron trichloride (BCl_3) and ammonia (NH_3) have the same number of atoms in a molecule, yet different shapes. Explain the shapes of these two molecules and why they are different from each other.

Question 9

Explain why some metallic elements produce flames of a distinctive colour when they are heated.

HIGHER-ORDER QUESTIONS

Question 1

a The elemental analysis of a compound showed that it contained carbon, hydrogen and nitrogen with percentage by mass of 54.6%, 13.6% and 31.8%, respectively. The compound was found to have a molecular mass of 88 g.

i Determine the empirical formula for this compound.

ii Determine the molecular formula for this compound.

b When hydrated salts are heated they release some or all of the water of crystallisation. Copper sulfate pentahydrate ($CuSO_4.5H_2O$) behaves in this manner. The mass of the product obtained when a 5.838 g sample of this compound was heated at 200°C until constant mass was achieved, was found to be 4.153 g.

i Determine the amount, in moles, of the original compound in the sample.

ii Determine the amount, in moles, of the water of crystallisation removed from the sample when it was heated to constant mass.

iii Use the ratio of the amount of the original compound to the amount of water removed on heating, to determine the formula for the product formed in this process.

iv Give one reason why the sample was heated until constant mass was achieved.

Question 2

a The first ionisation energies of the elements (i.e. the energy required to remove one electron from an atom) varies moving across a period and down a group in the periodic table.

i How does the first ionisation energy for the elements change moving across a period in the periodic table?

ii Briefly explain why this variation occurs moving across a period in the periodic table.

iii How does the first ionisation energy of the elements change moving down a group in the periodic table?

iv Briefly explain why this variation occurs moving down a group in the periodic table.

b Sodium, silicon and sulfur are three elements that are in the third period of the periodic table.

i Compare the metallic characteristics of sodium with those of sulfur.

ii What character would silicon be expected to display?

c Calculate the relative atomic mass of the element using the mass spectroscopy data in the following table and use your answer to identify the element.

Mass number	Percentage abundance
46	7.93
47	7.28
48	73.94
49	5.51
50	5.34

Question 3

a Aqueous solutions of hydrogen peroxide (H_2O_2) will decompose to form oxygen gas and water. Write an appropriate chemical equation for this reaction.

b The reaction in question **3a** is slow; however, when a catalyst is added, oxygen gas is readily formed. When a 50.0 mL sample of an aqueous hydrogen peroxide solution decomposed in the presence of a catalyst, 476 mL of dry oxygen gas at 22.6°C and 99.9 kPa was produced.

i Calculate the number of moles of oxygen gas formed.

ii Calculate the concentration of the hydrogen peroxide solution.

c Explain why the reaction above is faster when a catalyst has been added.

Question 4

The complete combustion of propane can be represented by the following chemical equation.

$$C_3H_8(g) + 5O_2(g) \rightarrow 3CO_2(g) + 4H_2O(g)$$

a A stoichiometric mixture of propane and oxygen was placed in a cylinder of fixed volume that was fitted with a spark plug. The initial pressure of the gases in the cylinder at 20°C was 254 kPa. The gas mixture was ignited using the spark plug. What would be the gas pressure inside the cylinder when the temperature returned to 20°C?

b An expandable helium balloon has a volume of 400 L when it was filled at 22°C and 100.6 kPa. The balloon was released into the atmosphere and at a certain altitude the temperature and pressure were recorded as 4.18 kPa and –40°C. What would be the volume of the balloon at this altitude?

Question 5

a The amount of energy required to convert a liquid into a gas is known as the latent heat of vaporisation. The latent heat of vaporisation for methane and water, two molecular compounds with comparable molar masses, are 8 kJ mol^{-1} and 44 kJ mol^{-1}, respectively. Explain why the latent heat of vaporisation for water is significantly higher than that for methane.

b The solubility of magnesium sulfate ($MgSO_4$) in water at 20°C is 36.2 g 100 mL^{-1}. A 500.0 mL sample of an aqueous magnesium sulfate solution at 20°C contained 159.5 g of magnesium sulfate.

i Determine the concentration of the magnesium ions in this solution.

ii Determine if this solution is unsaturated, saturated or supersaturated.

Question 6

Consider the following statements relating to exothermic and endothermic reactions.

A Exothermic reactions always start spontaneously whereas endothermic reactions require a source of energy to start.

B In exothermic reactions, the release of heat energy causes the products to have a lower chemical energy content than the reactants.

C On an energy profile for an endothermic reaction, the reactants have lower energy than the products.

D In the presence of a catalyst the energy required to break reactant bonds is decreased but the energy released when product bonds form is not affected.

E At a higher temperature, both exothermic and endothermic reactions speed up.

F The rate of an exothermic reaction increases as it proceeds but the rate of an endothermic reaction decreases as it proceeds.

a Identify two statements from this list (**A**–**F**) that are correct and for each one, provide an explanation to justify your response.

b Select two of the statements that are incorrect or partially correct and, for each one, explain why this is the case.

c Two students investigated the reaction between zinc and 2.0 mol L^{-1} hydrochloric acid. One used finely grained zinc powder, whereas the other one used lumps of zinc.

i Write a balanced equation for the reaction.

ii Explain, referring to collision theory, which reaction would be faster.

9780170412391

ANSWERS

CHAPTER 1 PERIODIC TABLE AND TRENDS

1.1 THE GEOGRAPHY OF THE PERIODIC TABLE

Students' own colour schemes to reflect the answers.

1.2 IMPORTANT TERMS

1 Amphoteric
2 Element
3 Period
4 Radius
5 Transition
6 Metalloid
7 Group
8 Noble
9 Alkali
10 Halogens
11 Electronegativity
12 Ionisation
13 Students' own representations.

1.3 EXPLAINING TRENDS IN ATOMIC RADIUS

1 Students' own representations.
- **a** Calcium has a greater number of electron shells than beryllium.
- **b** Nitrogen has a greater number of protons in the nucleus than beryllium. Therefore, there is a stronger electrostatic attraction between the nucleus and electrons in nitrogen and so the electrons are pulled more tightly towards the nucleus, causing the atom to be smaller.

1.4 EXPLAINING TRENDS IN ELECTRONEGATIVITY

1 Carbon has two electron shells, silicon has three. Therefore, carbon has a smaller radius so the electrostatic attraction of the nucleus to external electrons is greater. Hence carbon is more electronegative than silicon.

2 Oxygen has 8 protons in the nucleus, carbon only has 6. They have the same number of shells. Therefore, the electrostatic attraction of the nucleus to external electrons is greater for oxygen, so it is more electronegative.

1.5 EXPLAINING TRENDS IN FIRST IONISATION ENERGY

1 Students' own representations.

2 Boron has only two electron shells, whereas aluminium has three. Hence, the outermost electron in boron is closer to the nucleus than in aluminium, and therefore is more tightly bound, requiring more energy for it to be removed. Hence the first ionisation energy is greater.

3 Fluorine has a greater number of protons in the nucleus than boron, and the same number of electron shells. Therefore, the attraction between the nucleus and the outermost electron is greater for fluorine than for boron, so the first ionisation energy is greater also.

1.6 OXIDES

1 D – aluminium oxide is described as amphoteric
2 C – phosphorous is a non-metal
3 C – magnesium is a metal – an alkaline earth metal

EVALUATION

1 **a** Alkali metals
- **b** **i** Amount of energy required to remove the outermost electron from a gaseous atom
  - **ii** Ability of a bonded atom to attract an electron
- **c** Lithium has fewer electron shells than potassium
- **d** First ionisation of sodium would be less than that of lithium but greater than that of potassium. Ionisation energy decreases as the number of electron shells increases, and the outer electrons become further away from the nucleus and so are less strongly attracted to it.
- **e** Electronegativity of beryllium might be around 1.00 – larger than that of lithium, as electronegativity increases across a period, due to the increased number of protons in the nucleus, and so higher attraction to external electrons.

CHAPTER 2 ATOMIC STRUCTURE

2.1 IMPORTANT TERMS

Across

1 Nucleus
3 Atomic
5 Strong
6 Configuration
7 Aufbau
9 Isotopes
11 Element
12 Orbitals
13 Atoms
14 Shell

Down

2 Electrostatic
4 Electron
6 Charged
8 Proton
10 Neutrons

2.2 NUMBERS OF PROTONS, NEUTRONS AND ELECTRONS

Complete the following table to show the correct information for the atoms and ions listed:

SYMBOL	CHARGE	ATOMIC NUMBER	MASS NUMBER	NUMBER OF PROTONS	NUMBER OF NEUTRONS	NUMBER OF ELECTRONS
Li	0	3	7	3	4	3
O	2–	8	17	8	9	10
Mg	2+	12	24	12	12	10
As	3–	33	75	33	42	36
B	3+	5	7	5	2	2
Kr	0	36	83	36	47	36
F	1–	9	19	9	10	10
Cl	0	17	37	17	20	17
Ca	2+	20	41	20	21	18
Mg	0	12	26	12	14	12
Cu	2+	29	62	29	33	27
Cu	1+	29	63	29	34	28

2.3 ELECTRON CONFIGURATIONS

1 Electrons fill shells, from the lowest energy (innermost) to the highest energy (outermost)

2 1s, 2s, 2p, 3s, 3p, 4s, 3d, 4p

3

PERIOD 1	PERIOD 2	PERIOD 3	PERIOD 4
H = $1s^1$ He =$1s^2$	Li = [He] $2s^1$ Be [He] $2s^2$ B = [He] $2s^2$ $2p^1$ C = [He] $2s^2$ $2p^2$ N = [He] $2s^2$ $2p^3$ O = [He] $2s^2$ $2p^4$ F = [He] $2s^2$ $2p^5$ Ne = [He] $2s^2$ $2p^6$	Na = [Ne] $3s^1$ Mg = [Ne] $3s^2$ Al = [Ne] $3s^2$ $3p^1$ Si = [Ne] $3s^2$ $3p^2$ P = [Ne] $3s^2$ $3p^3$ S = [Ne] $3s^2$ $3p^4$ Cl = [Ne] $3s^2$ $3p^5$ Ar = [Ne] $3s^2$ $3p^6$	K = [Ar] $4s^1$ Ca = [Ar] $4s^2$ Sc = [Ar] $4s^2$ $3d^1$ Ti = [Ar] $4s^2$ $3d^2$ V = [Ar] $4s^2$ $3d^3$ Cr = [Ar] $4s^1$ $3d^5$ Mn = [Ar] $4s^2$ $3d^5$ Fe = [Ar] $4s^2$ $3d^6$ Co = [Ar] $4s^2$ $3d^7$ Ni = [Ar] $4s^2$ $3d^8$ Cu = [Ar] $4s^1$ $3d^{10}$ Zn = [Ar] $4s^2$ $3d^{10}$ Ga = [Ar] $4s^2$ $3d^{10}$ $4p^1$ Ge = [Ar] $4s^2$ $3d^{10}$ $4p^2$ As = [Ar] $4s^2$ $3d^{10}$ $4p^3$ Se = [Ar] $4s^2$ $3d^{10}$ $4p^4$ Br = [Ar] $4s^2$ $3d^{10}$ $4p^5$ Kr = [Ar] $4s^2$ $3d^{10}$ $4p^6$

4 **a** Hydrogen H^+ $1s^0$

b Beryllium Be^{2+} $1s^2$

c Nitrogen N^{3-} [He] $2s^2$ $2p^6$

d Aluminium Al^{3+} [Ne] $3s^0$

e Sulfur S^{2-} [Ne] $3s^2$ $3p^6$

f Potassium K^+ [Ar] $4s^0$

g Scandium Sc^{3+} [Ar] $4s^0$

h Vanadium V^{2+} [Ar] $4s^0$ $3d^3$

i Manganese (II) Mn^{2+} [Ar] $4s^0$ $3d^5$

j Manganese (III) Mn^{3+} [Ar] $4s^0$ $3d^4$

k Nickel (II) Ni^{2+} [Ar] $4s^0$ $3d^8$

l Copper (I) Cu^{1+} [Ar] $4s^0$ $3d^{10}$

m Copper (II) Cu^{2+} [Ar] $4s^0$ $3d^9$

n Germanium (IV) Ge^{4+} [Ar] $4s^0$ $3d^{10}$ $4p^0$

o Bromide Br^{1-} [Ar] $4s^2$ $3d^{10}$ $4p^6$

2.4 KEY IDEAS

1 a All matter is composed of extremely small particles called **atoms**.

b Atoms have an internal structure consisting of a tiny, dense central area called the **nucleus**, surrounded largely by empty space.

c Atoms are formed out of three types of subatomic particles: **protons**, **neutrons** and **electrons**.

d Protons and neutrons are **two** subatomic particles found inside the nucleus.

e Electrons are **negative** subatomic particles found outside of the nucleus.

f Protons and neutrons, which are densely packed together in the nucleus, provide most of an atom's **mass**.

g Electrons, which move around the nucleus at high **speed**, occupy most of the volume of the atom but provide little of its mass.

h A proton has one **positive** charge and an electron has one **negative** charge, but a neutron is **neutral**.

i There are 92 different types of naturally occurring atoms. Each type is defined by the number of **protons** in its nucleus. For example, every carbon atom has six of these subatomic particles.

j Uncharged atoms have **equal** numbers of protons and electrons, so an uncharged carbon atom will have **six** electrons.

k A substance consisting of only one type of atom is called an **element**. For example, the element carbon consists of atoms with six **protons** only.

l Each element is given a unique name and **symbol** (e.g. magnesium, Mg).

m Each type of atom can have varying numbers of **neutrons** in the nucleus. For example, carbon atoms can have 6, 7 or even 8 of them. These different forms of an atom are known as **isotopes**.

n An atom's electrons have discrete amounts of energy and are located in specific energy levels, known as **shells** around the nucleus, depending on how much energy they have. Those in the innermost shells have the **lowest** amount of energy.

o Each shell can accommodate up to a **maximum** number of electrons. The first shell accommodates up to **two**, the second shell **eight** and the third shell up to **eighteen** each.

p Each shell contains a series of **orbitals**, which can be labelled s, p, d or f.

q The **Aufbau** principle states that the electron shells are filled, lowest in energy first.

r The Pauli **exclusion** principle states that each electron has a unique location within the atom, which can be represented using the s, p, d, f notation.

s When ions are formed, electrons are lost from the **highest** energy level first.

t The periodic table can be divided into **blocks**, according to the subshell in which the outermost electron exists.

EVALUATION

1 a Protons are found within the nucleus, and are positively charged. Neutrons are also found in the nucleus, are uncharged and have a mass equal to that of a proton. Electrons are found in energy levels around the nucleus. They are negatively charged and have a negligible mass compared to protons and neutrons.

b i Positively charged protons attract negatively charged electrons and hold them to the nucleus

ii Protons and neutrons are held together in the nucleus, to ensure it is stable and does not decay

2 a

SPECIES	NUMBER OF PROTONS	NUMBER OF NEUTRONS	NUMBER OF ELECTRONS
${}^{49}_{22}X$	22	27	22
${}^{42}_{20}Z^{2+}$	20	22	18

b $1s^2\ 2s^2\ 2p^6\ 3s^2\ 3p^6\ 4s^2\ 3d^2$

c It has lost two electrons, so has two more protons than electrons

3 a i Ti: [Ar] $4s^2\ 3d^2$

ii V: [Ar] $4s^2\ 3d^3$

iii Cr: [Ar] $4s^1\ 3d^5$

b Additional stability of the half-filled d orbital means that $4s^1\ 3d^5$ is a lower energy configuration than the expected $4s^2\ 3d^4$.

CHAPTER 3 INTRODUCTION TO BONDING

3.1 IMPORTANT TERMS

Across

2 Lone

4 Compound

9 Cation

11 Electron

13 Valency

14 Ion

Down

1 Anion

3 Ionic

5 Delocalised

6 Covalent

7 Metallic

8 Double

10 Valence

12 Electrostatic

3.2 COMPARISON OF BONDING TYPES

BONDING MODELS	DISTINCTIVE FEATURES	COMMON FEATURES
Electron cloud (negative)	• covalent bonding • shared pair of electrons • between non-metal atoms • no charged particles	• all bonding involves electrons • all involves electrostatic attraction
Key: Sodium ion Mobile valence electron	• metallic bonding • positive ions surrounded by 7e localised electrons • no compound formed	
Na^+ Cl^-	• ionic bonding • electron transfer • formation of positive and negative ions • between metal and non-metal atom	

3.3 KEY IDEAS

a	Anion	O
b	Bonding pair	Q
c	Cation	R
d	Chemical bond	C
e	Compound	S
f	Covalent bonding	D
g	Molecular compound	L
h	Delocalised electrons	J
i	Directional bonding	K
j	Double covalent bond	H
k	Ion	F
l	Ionic bonding	A
m	Ionic compound	G
n	Metallic bonding	B
o	Multiple covalent bond	I
p	Non-bonding (lone) pair	Q
q	Non-directional bonding	E
r	Polyatomic ion	N
s	Valence electrons	M
t	Valency	T

3.4 IONIC AND COVALENT FORMULAS

1 a $NaOH$

b $Ca(NO_3)_2$

c $ZnSO_4$

d Co_2O_3

e $Al_2(CO_3)_3$

f $Cu_3(PO_4)$

2 a SiF_4

b CO_2

c CO

d $SeCl_2$

e Cl_2O

f NBr_3

9780170412391

EVALUATION

1

NAME	MOLECULAR FORMULA	VALENCE STRUCTURE
Carbon tetrachloride	CCl_4	Cl–C(Cl)(Cl)–Cl
Carbon dioxide	CO_2	O=C=O
Boron trichloride	BCl_3	Cl–B(Cl)–Cl
Ammonia	NH_3	H–N(H)–H

2

NAME	FORMULA	BONDING TYPE
Sodium oxide	Na_2O	ionic
Sulfur dioxide	SO_2	covalent
Boron sulfide	B_2S_3	covalent
Manganese (II) chloride	$MnCl_2$	ionic
Selenium iodide	SeI_2	covalent

3 a The magnesium atom loses two valence electrons to form a Mg^{2+} ion. The oxygen atom gains two electrons to form an O^{2-} ion. There is an attraction between the two ions to form magnesium oxide, a new compound.

b When magnesium mixes with sodium, the metallic bonding structure remains – the sodium and magnesium ions form a lattice and delocalised electrons from both elements mix. This creates an alloy, rather than a new substance and the properties of the alloy vary according to the proportions in which the two metals are mixed.

4 Phosphorous has valence electrons. Three of these form covalent bonds by being shared with electrons from the chlorine atoms, leaving two unpaired electrons – the lone pair.

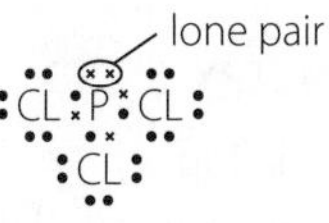

CHAPTER 4 ISOTOPES

4.1 SMOKE DETECTORS

1 An americium atom decays to form a neptunium atom and an alpha particle, which is a Helium nucleus.

2 Radioactive material is present.

3 $^{241}_{95}Am$ has 95 protons and 146 neutrons. $^{237}_{93}Np$ has 93 protons and 144 neutrons.

4 An alpha particle consists of 2 protons and 2 neutrons – the difference between an Am and a Np atom.

5 They ionise them, causing them to become charged.

6 A battery is required to maintain an electric field across the chamber.

7 The charged particles can move and carry the current.

8 Ionised smoke particles reduce the ionisation current because they move more slowly than air particles.

9 Continued smoke production – possibly caused by a fire.

10 They cannot travel more than a few centimetres, so will not reach a human

4.2 ISOTOPE OR NOT?

1

ATOM	ATOMIC NUMBER	MASS NUMBER	NUMBER OF PROTONS	NUMBER OF NEUTRONS	NUMBER OF ELECTRONS
Z	7	16	7	9	7
Y	12	28	12	16	12
X	18	34	18	16	18
W	7	18	7	11	7
V	11	25	11	14	11
U	7	15	7	8	7
T	11	22	11	11	11

2 Atoms U, W and Z are isotopes of each other.

Atoms T and V are isotopes of each other.

4.3 RELATIVE ATOMIC MASS CALCULATIONS

1 $RAM = (0.01 \times 112) + (0.007 \times 114) + (0.003 \times 115) + (0.145 \times 116) + (0.077 \times 117) + (0.242 \times 118) + (0.086 \times 119) + (0.326 \times 120) + (0.046 \times 122) + (0.058 \times 124) = 118.8$

2 $30.85 = (30 \times a) + (33 \times (1 - a))$

$30.85 - 33 = 30a - 33a$

$3a = 2.15$

$a = 0.72$

% abundance of R-30 is 72% and of R-33 is 28%

EVALUATION

1 Two atoms of the same element with different numbers of neutrons.

2 The average mass of an atom of the element, relative to carbon-12

3 85% Y-33 and 15% Y-34

4 a They have the same electron configuration and therefore the same chemical properties. Different numbers of neutrons in the nucleus can affect the strong forces holding the protons and neutrons together and, therefore, the nuclei of some isotopes can be less stable and prone to decay

b Carbon-14 could form carbon dioxide which reacted in exactly the same way as that made from carbon-12. However, the reactive carbon-14 could be detected and it is quite rare, so its progress can be followed through the reaction pathway.

CHAPTER 5 ANALYTICAL TECHNIQUES

5.1 IMPORTANT TERMS

Across

1 Quantitative

2 AAS

4 Photon

8 Spectrometry

Down

1 Qualitative

3 Absorption

5 Emission

6 Bohr

7 Flame

5.2 CONSTRUCTING AND USING CALIBRATION CURVES

1

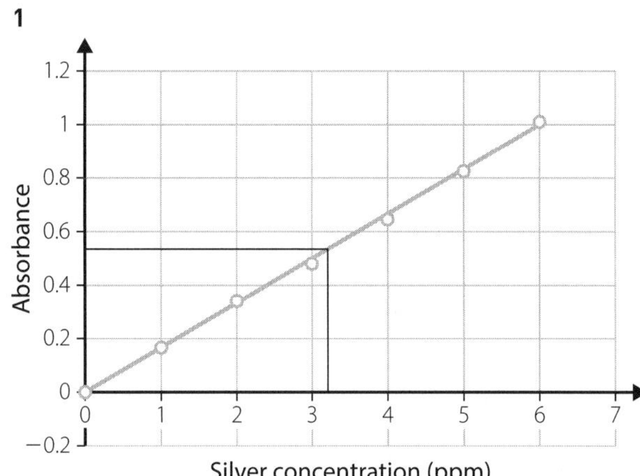

2 3.2 ppm (see red line)

3

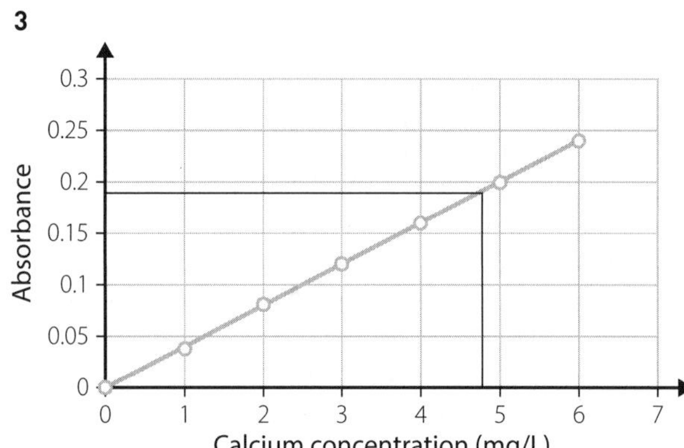

4 4.8 mg/L

5 The filament of the lamp is made from the metal to be investigated, so to adjust the machine, replace a lamp with a lead filament for one with a cadmium filament.

5.3 MASS SPECTROMETRY

1 Numbers of different isotopes present, the relative mass of the isotopes and the relative abundance of each isotope.

2 24.3

9780170412391

5.4 EMISSION AND ABSORPTION SPECTROSCOPY

1 The excited state occurs when energy has been absorbed, causing an electron to move to a higher energy level. Eventually, the additional energy will be lost and the electron will fall back to its original ground state.

2 It moves into a higher energy level provided it absorbs the specific frequency of energy corresponding to the energy difference between the levels.

3 The frequencies of light emitted by electrons when they have absorbed light and are falling back to their original energy level – the ground state.

4 The two elements have different electron configurations and the specific values of the energy levels changes when the number of electrons in them changes. Hence the frequencies emitted by two different elements will be distinct.

5 Students' own diagrams.

EVALUATION

1 a

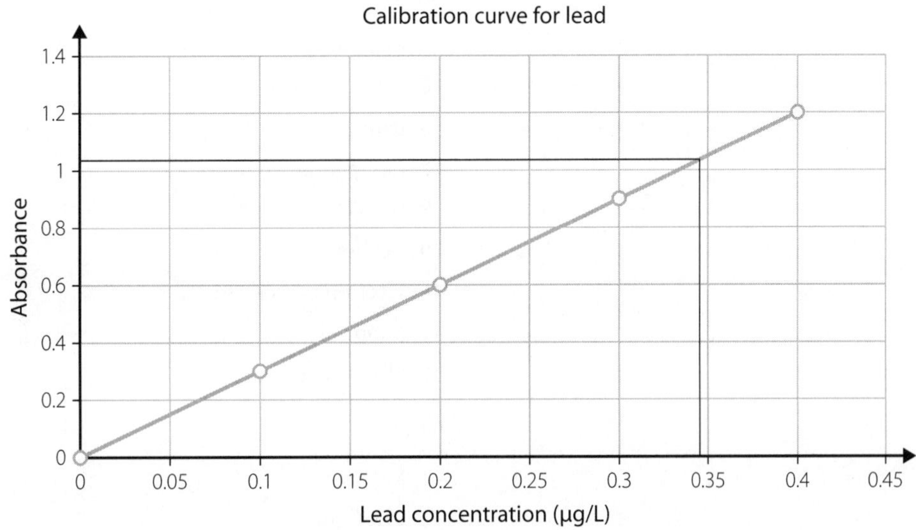

Concentration of lead in Solution B is: 0.345 µg/L

b Concentration in Solution A = 250/1 × 0.345 = 86.25 µg/L

c Original mass of lead is 500/1000 × 86.25 = 43.10.31425 µg = 0.043 mg

Concentration of lead in fish = 0.043 / 0.1373 = 0.314 mg/g = 314 mg/kg

d This fish should not be sold for human consumption – the level is significantly above the recommended maximum.

e It uses a lamp with the element made from lead. Hence only lead frequencies are emitted and will be absorbed selectively by the lead ions. Other metal ions present will not be able to be absorbed.

CHAPTER 6 COMPOUNDS AND MIXTURES

6.1 IMPORTANT TERMS

1 a The **solid** state of matter has a defined volume and shape.

b The **liquid** state of matter has a defined volume, but not a defined shape.

c The **gas** state of matter has neither a defined volume, nor a defined shape.

d Substances that are found in nature are called **natural**.

e Substances that are created in a laboratory are called **synthetic**.

f A **physical** process starts and finishes with the same substances.

g A **chemical** process starts and finishes with different substances.

h A **microscope** enables scientists to view phenomena too small to be seen with the naked eye.

i **Heating** converts a substance in solid form into liquid form.

j **Boiling** converts a substance in liquid form into gas form.

k **Solubility** is a measure of a solute's ability to dissolve in a solvent.

l **Flammability** is a measure of a substance's ability to burn.

m In meteorology, **precipitation** means rain.

n In chemistry, **precipitation** means a solid is formed.

o In biology, **evolution** means a change in heritable characteristics over successive generations.

p In chemistry, **evolution** means a gas is produced.

q **Temperature** is a quantitative measure of a substance's hotness or coldness.

r **Melting** is increasing the temperature of a substance.

s **Heat** energy is a function of temperature, mass and the type of substance.

t **Desalination** converts salt water into fresh water.

u **Uniformity** is a measure of a substance's sameness of composition throughout.

2 a L

b S

c T

d C

e G

f F

g B

h Q

i R

j K

k J

l H

m N

n P

o O

p A

q D

r E

s M

t I

3 a **Diesel**: a mixture of hydrocarbon chains containing 10–24 carbon atoms. The most common component is hexadecane, $C_{16}H_{34}$; **Petrol**: a mixture of hydrocarbons containing 5–12 carbon atoms. The most common component is octane, C_8H_{18}

b **Hydrophilic**: 'water-loving', a particle with polar regions that bond to water; **Hydrophobic**: 'water-hating', a particle with mostly non-polar regions that do not bond with water

c **Natural diamonds**: natural diamonds are formed from carbon-containing minerals at high temperature and pressure at depths of 140–190 km in Earth's mantle; **Synthetic diamonds**: Synthetic diamonds are diamonds manufactured in a laboratory usually by exposing carbon to high temperatures and high pressures

d **n-type region of a solar panel**: an area in which one silicon atom in the silicon lattice is replaced by a phosphorous atom; **p-type region of a solar panel**: an area in which one silicon atom in the silicon lattice is replaced by a boron atom

4

Across

1 Macroscopic

5 Solvent

6 Potash

7 Compound

9 OTWO

10 Helium

14 Electron microscope

16 Nano

17 Electrolysis

Down

1 Methane

2 CHFOUR

3 Optical microscope

4 Petrol

5 Solute

8 Davy

11 Conductivity

12 Quicklime

13 NaCl

15 COTWO

16 NOTWO

6.2 SUBSTANCES

1

SUBSTANCE	PURITY	HOMOGENEITY	COMPOSITION	SEPARATION
Element	Pure	Homogeneous	Fixed	Small
Compound	Pure	Homogeneous	Fixed	Small
Mixture	Impure	Heterogeneous	Variable	Large

9780170412391

2

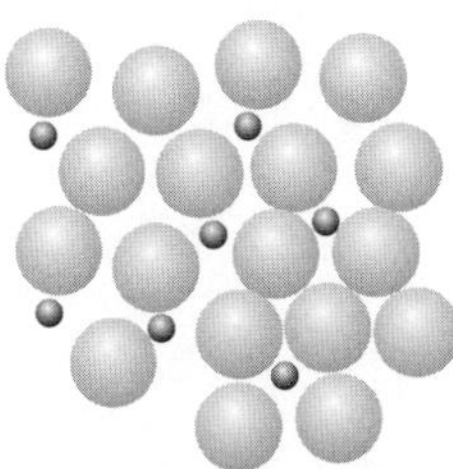

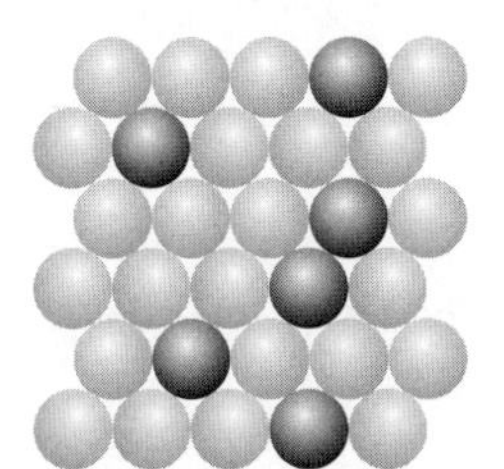

3 Salt is spread on icy roads to lower the freezing point of water. This prevents rain or snow from freezing. Antifreeze is placed in the radiators of car engines to lower the freezing point of water, which prevents the water in the radiator from freezing, expanding and causing damage to the radiator.

4 a Brown-orange gas produced

b Conduct experiment in a fume cupboard because of evolution of toxic brown–orange gas

c $Pb(NO_3)_2$ – lead (ii) nitrate

PbO – lead (ii) oxide

NO_2 – nitrogen dioxide

d O_2 – oxygen

e heterogeneous

5 a C_8H_{18} – octane

CO_2 – carbon dioxide

H_2O – water

b This reaction could be used as the energy source in internal combustion engines.

c The amount of heat produced

d More unburnt hydrocarbon emission

e Large quantities of CO_2 are produced, which contributes to global warming.

f i $CH_4(g) + 2O_2(g) \rightarrow CO_2(g) + 2H_2O(l)$

ii $2C_{16}H_{34}(l) + 49O_2(g) \rightarrow 32CO_2(g) + 34H_2O(l)$

6.3 SEPARATION METHODS

1 a Gravity filtration

b Solubility

c Table salt and sand

d Solvent extraction

e Immiscibility of two liquids

f Caffeine from coffee

g Vaporisation

h Low boiling point of solvent

i Salt and water

j Distillation

k Difference in boiling points

l Ethanol and water

6.4 REAL WORLD DISTILLATION

Students' answers will vary.

6.5 REAL WORLD SEPARATORY FUNNEL USE

Students' answers will vary.

6.6 REAL WORLD LIQUID–LIQUID EXTRACTIONS

Students' answers will vary.

EVALUATION

1

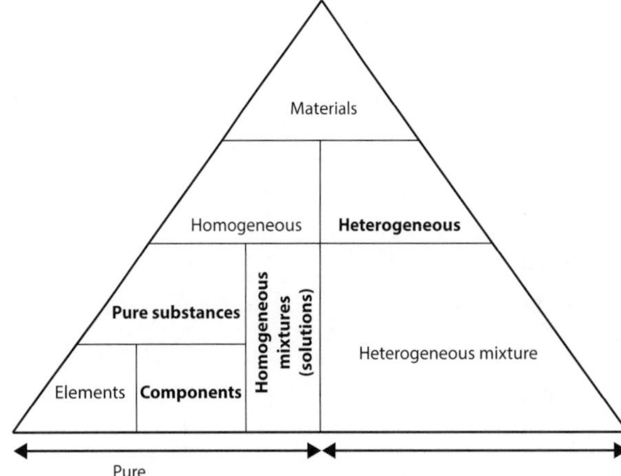

2 a Precipitate formed of $Cu(OH)_2(s)$; disappearance of blue $Cu^{2+}(aq)$ solution

b heterogeneous

c $Cu(OH)_2$

d filtration

3 a Black solid, $CuO(s)$ disappears; blue solution, $Cu^{2+}(aq)$ appears

b homogeneous

c CuO

d H_2O

4 a Vitamin D2

i Vitamin D2 is fat-soluble because of its long hydrocarbon-rich chain which can easily form intermolecular interactions with similarly long hydrocarbon-rich fat molecules.

ii Its long, hydrocarbon-rich chain

iii No. Being fat-soluble, it stays in the body longer.

b Vitamin B2

i The presence of a number of -OH groups concentrated at one end of the molecule impart its hydrophilicity due to the formation of strong hydrogen bonds with water molecule.

ii The carbon chain containing the four -OH groups

iii Yes. Due to its water solubility, it is excreted regularly and must be replaced daily.

c Answers will vary

d The main selling point would be that vitamin D2, being fat-soluble would not need to be consumed on a daily basis.

5 **a** High brass is a substitution alloy. Its components, copper and zinc have atomic radii that are within 15% of each other.

b Carbon steel is an interstitial alloy. Iron has a far greater atomic radius than carbon.

c Stainless steel is a substitutional and an interstitial alloy. The iron and chromium have the same atomic radii but the atomic radius of carbon is much smaller.

CHAPTER 7 BONDING AND PROPERTIES

7.1 IMPORTANT TERMS

Across

8 Malleable
11 Ductile
12 Saturated
13 Aromatic
14 Tensile
15 Density
16 Addition
17 Structural

Down

1 Substituent
2 Delocalised
3 Intramolecular
4 Replacement
5 Organic
6 Hardness
7 Unsaturated
9 Hydrocarbons
10 Crystal
18 Combustion
19 Allotrope

7.2 BONDING WITHIN SOLIDS

1

SOLID TYPE	EXAMPLE	DESCRIPTION OF STRUCTURE
Ionic	Sodium chloride	Crystal lattice
Covalent molecular	Methane	Discrete molecules
Covalent network	Diamond	Giant lattice
Covalent layer	Graphite	Giant molecular
Metallic	Iron	Lattice of metal ions

7.3 DRAWING AND NAMING HYDROCARBONS

1 **a** Pentane

```
    H   H   H   H   H
    |   |   |   |   |
H — C — C — C — C — C — H
    |   |   |   |   |
    H   H   H   H   H
```

b 2, 3 – Pentadiene

```
H             H
  \          /
   C = C = C
  /          \
H             CH2CH3
```

c Benzene

```
        H
        |
  H     C     H
    \ //  \  /
      C     C
      |     ||
      C     C
    /  \\  /  \
  H     C     H
        |
        H
```

d 3-Hexene

```
    H   H           H   H
    |   |           |   |
H — C — C — C = C — C — C — H
    |   |   |   |   |   |
    H   H   H   H   H   H
```

e 2-Methylbutane

```
            H
            |
        H — C — H
    H       |       H   H
    |       |       |   |
H — C ————— C ————— C — C — H
    |       |       |   |
    H       H       H   H
```

f 2, 2, 4-Trimethylpentane

```
                 H                H
                 |                |
             H — C — H        H — C — H
    H   H        |                |       H
    |   |        |                |       |
H — C — C ————— C ——————————————— C ————— C — H
    |   |        |                |       |
    H   H        |                H       H
             H — C — H
                 |
                 H
```

2 **a** 1-Pentene

```
            H   H   H
            |   |   |
H — C = C — C — C — C — H
    |   |   |   |   |
    H   H   H   H   H
```


9780170412391

b 2-Pentene

```
      H           H   H
      |           |   |
  H — C — C = C — C — C — H
      |   |   |   |   |
      H   H   H   H   H
```

c 2-Methyl-1-butene

```
                 H   H
                 |   |
  H — C = C  —   C — C — H
      |   |      |   |
      H   |      H   H
      H — C — H
          |
          H
```

d 3-Methyl-1-butene

```
                  H     H
                  |     |
  H — C = C  —    C  —  C — H
      |   |       |     |
      H   H       |     H
              H — C — H
                  |
                  H
```

e 2-Methyl-2-butene

```
      H                 H
      |                 |
  H — C  —  C  =  C  —  C — H
      |     |     |     |
      H     |     H     H
        H — C — H
            |
            H
```

7.4 REACTIONS OF HYDROCARBONS

1 **a** $2C_6H_{14} + 19O_2 \rightarrow 12CO_2 + 14H_2O$

b $CH_2C(CH_3)CH_2CH_2CH_3 + H_2 \rightarrow CH_3CH(CH_3)CH_2CH_2CH_3$

c $CH_3CHCHCH_3 + H_2O \rightarrow CH_3CHOHCH_2CH_3$

d $C_6H_6 + Br_2 \rightarrow C_6H_5Br + HBr$

e $CH_3CHCHCH_2CH_3 + Br_2 \rightarrow CH_3CHBrCHBrCH_2CH_3$

EVALUATION

1 A

2

DESCRIPTION	MELTING POINT	CONDUCTIVITY WHEN SOLID	CONDUCTIVITY WHEN MOLTEN	COVALENT MOLECULAR, IONIC OR METALLIC?
Silvery solid	98°C	Yes	Yes	Metallic
White solid	920°C	No	Yes	Ionic
Shiny solid	Sublimes (vaporises immediately on forming a liquid)	No	–	Covalent molecular
Yellow solid	110°C	No	No	Covalent molecular

3 **a** **Allotrope**: a different physical form of the same element e.g. carbon(diamond), carbon(graphite); **Isotope**: a different form of an element with the same number of protons but a different number of neutrons e.g. ^{12}C (6 protons and 6 neutrons), ^{13}C (6 protons and 7 neutrons)

b **Covalent molecular structure**: discrete molecules in which the atoms are held together by covalent bonding, the molecules being held together by weak intermolecular forces; **Covalent network structure**: a three-dimensional network of covalently bonded atoms

4 **a** Hexan-1-ol

b 2, 2-Dibromopentane

c butane

CHAPTER 8 CHEMICAL REACTIONS

8.1 CHEMICAL REACTIONS

1 Physical

2 Chemical

3 Chemical

4 Physical

5 Chemical

6 Chemical

7 Physical

8 Physical

9 Chemical

10 Physical

11 Chemical

12 Chemical

8.2 BALANCING CHEMICAL EQUATIONS

1 **a** aluminium + oxygen → aluminium oxide

$Al(s) + O_2(g) \rightarrow Al_2O_3(s)$

$4Al(s) + 3O_2(g) \rightarrow 2Al_2O_3(s)$

b hydrogen + oxygen → water

$H_2(g) + O_2(g) \rightarrow H_2O(l)$

$2H_2(g) + O_2(g) \rightarrow 2H_2O(l)$

c zinc + hydrochloric acid → zinc chloride + hydrogen

$Zn(s) + HCl(aq) \rightarrow ZnCl_2(aq) + H_2(g)$

$Zn(s) + 2HCl(aq) \rightarrow ZnCl_2(aq) + H_2(g)$

d potassium hydroxide + sulfuric acid → potassium sulfate + water

$KOH(aq) + H_2SO_4(aq) \rightarrow K_2SO_4(aq) + H_2O(l)$

$2KOH(aq) + H_2SO_4(aq) \rightarrow K_2SO_4(aq) + H_2O(l)$

e magnesium chloride + sodium hydroxide → magnesium hydroxide + sodium chloride

$MgCl_2(aq) + NaOH(aq) \rightarrow NaCl(aq) + Mg(OH)_2(s)$

$MgCl_2(aq) + 2NaOH(aq) \rightarrow 2NaCl(aq) + Mg(OH)_2(s)$

f barium nitrate + potassium sulfate → potassium nitrate + barium sulfate

$Ba(NO_3)_2(aq) + K_2SO_4(aq) \rightarrow KNO_3(aq) + BaSO_4(s)$

$Ba(NO_3)_2(aq) + K_2SO_4(aq) \rightarrow 2KNO_3(aq) + BaSO_4(s)$

g nitrogen + hydrogen → ammonia

$N_2(g) + H_2(g) \rightarrow NH_3(g)$

$N_2(g) + 3H_2(g) \rightarrow 2NH_3(g)$

2 a $2\,Li + I_2 \rightarrow 2\,LiI$

b $2\,NaClO_3 \rightarrow 2NaCl + 3O_2$

c $K_3PO_4 + 3HCl \rightarrow 3KCl + H_3PO_4$

d $C_5H_{12} + 8O_2 \rightarrow 5CO_2 + 6H_2O$

e $3Ca(NO_3)_2 + Na_3PO_4 \rightarrow Ca_3(PO_4)_2 + 6NaNO_3$

f $C_6H_8 + 8O_2 \rightarrow 6CO_2 + 4H_2O$

g $H_3PO_4 + 3KOH \rightarrow K_3PO_4 + 3H_2O$

h $2Al(OH)_3 + 3H_2CO_3 \rightarrow Al_2(CO_3)_3 + 6H_2O$

i $4\,FeS_2 + 1O_2 \rightarrow 2Fe_2O_3 + 8SO_2$

j $Ca_3(PO_4)_2 + 3SiO_2 + 5C \rightarrow 3CaSiO_3 + 5CO + 2\,P$

8.3 CHEMICAL REACTIVITY

1

	SODIUM	MAGNESIUM	ZINC	COPPER
Dropped into water	Reacts vigorously to yield sodium hydroxide and hydrogen gas	Reacts very slowly to yield magnesium hydroxide and small bubbles of hydrogen gas. The reaction soon stops because magnesium hydroxide forms an insoluble coating on the magnesium preventing further reaction	Undergoes virtually no reaction (Note that zinc reacts with steam to yield zinc oxide and hydrogen gas)	No reaction
Heated	Burns with a yellow flame to yield sodium oxide (Na_2O) and sodium peroxide (NA_2O_2)	Burns with a bright white flame to yield a white powder of magnesium oxide (MgO). Also reacts with N_2 to form magnesium nitride.	Heated strongly to high temperatures in air burns to yield zinc oxide (white powder).	Heated strongly in air to very high temperatures turns black due to the formation of copper oxide.
Added to dilute hydrochloric acid	Reacts explosively to form sodium chloride (NaCl) and hydrogen gas	Reacts vigorously to form magnesium chloride ($MgCl_2$) and a burst of hydrogen gas	Reacts slowly to form zinc chloride $ZnCl_2$	No reaction

2 Copper is the only metal suitable for making saucepans. Cooking involves heat and usually water, which would lead to reactions of Na, Mg and Zn to form oxides and hydroxides. This would degrade the utensil quickly. Foods often contain acids and all the metals other than copper react with acids, sodium explosively.

3 a Yes

b No

c Yes

d Yes

4 a F

b T

c F

d T

e T

f T

5 a $Zn(s) + 2H_2O(g) \rightarrow ZnO(s) + H_2(g)$

b $2Na(s) + 2HCl(aq) \rightarrow 2NaCl(aq) + H_2(g)$

c $2Mg(s) + O_2(g) \rightarrow 2MgO(s)$

6 a $Zn(s) + Cu^{2+}(aq) \rightarrow Cu(s) + Zn^{2+}(aq)$

Zn lies higher in the metal activity series than Cu hence it will displace Cu.

b $CO_3^{2-}(aq) + Ca^{2+}(aq) \rightarrow 2CaCO_3(s)$

Calcium carbonate is relatively insoluble, hence it precipitates from solution.

7 a $Mg(s) + 2H^+(g) + 2OH^-(g) \rightarrow Mg(OH)_2(s) + H_2(g)$

b $Zn(s) + 2H^+(aq) \rightarrow Zn^{2+}(aq) + H_2(g)$

c $Fe(s) + 2H^+(aq) \rightarrow Fe^{2+}(aq) + H_2(g)$ (formation of ferrous or Fe(II) ions)

also:

$2Fe(s) + 6H^+(aq) \rightarrow 2Fe^{3+}(aq) + 3H_2(g)$ (formation of ferric or Fe(III) ions)

d $Pb^{2+}(aq) + 2I^-(aq) \rightarrow PbI_2(s)$

9780170412391

EVALUATION

1 a Physical – no bonds are broken or new bonds created, no heat, light released.

b Chemical – reaction of organic material (wood) with oxygen yields CO_2 – bonds are broken, new bonds are formed; heat is released.

c Physical – no change in chemical structure – only a change of state from liquid phase to gas phase.

d Chemical – reaction occurs, evidenced by the production of hydrogen gas.

e Physical – no bonds are broken or new bonds formed solid is dissolved hence it is a transformation from solid phase to aqueous phase, i.e. a phase change.

f Chemical – heat is produced (explosion); reaction of H_2 with O_2 yields H_2O; bonds are broken, new bonds are formed.

2 a $2N_2 + O_2 \rightarrow 2N_2O$

b $2NaI + Cl_2 \rightarrow 2NaCl + I_2$

c $C_3H_8 + 5O_2 \rightarrow 3CO_2 + 4H_2O$

d $3Mg(NO_3)_2 + 2K_3PO_4 \rightarrow Mg_3(PO_4)_2 + 6KNO_3$

3 c Gold

4 a $2C_6H_{14}(aq) + 19O_2(g) \rightarrow 12CO_2(g) + 14H_2O(g)$

b $2Fe(s) + O_2(aq) + 2H_2O(aq) \rightarrow 2Fe(OH)_2(aq)$

c $N_2(g) + 3H_2(g) \rightarrow 2NH_3(g)$

CHAPTER 9 EXOTHERMIC AND ENDOTHERMIC REACTIONS

9.1 IMPORTANT TERMS

Across

4 Negative
5 Temperature
6 Endothermic
10 Hess
11 Vibrational
13 Rotational

Down

1 Positive
2 Energy
3 Combustion
5 Translational
7 Exothermic
8 Kelvin
9 Enthalpy
12 Celsius

9.2 EXOTHERMIC AND ENDOTHERMIC REACTIONS

PROCESS	EXO OR ENDO	PROCESS	EXO OR ENDO
Water vapour condensing to liquid water (e.g. clouds turning to rain)	Exothermic	Conversion of frost to water vapour	Endothermic
Producing sugar and O_2 by photosynthesis	Endothermic	Burning gas or petrol (combustion)	Exothermic
Mixing sodium hydroxide pellets with water	Exothermic	Mixing water with ammonium nitrate	Endothermic
Mixing water with strong acids	Exothermic	Melting of ice cubes or snow	Endothermic
Splitting apart a gas molecule (e.g. O_2 or N_2)	Endothermic	Mixing water with cooking salt	Endothermic
Snow formation in clouds	Exothermic	Baking bread	Endothermic
Burning sugar	Exothermic	Cooking an egg	Endothermic
Nuclear fission	Exothermic	Making ice cubes in a freezer	Exothermic
The evaporation of water	Endothermic	Rusting of iron in air	Exothermic

2 a Bonds broken: C—H (all four bonds in CH_4) and O—O (one bond in both O_2 molecules)

Bonds formed: C—O (two C—O bonds) and O—H (two O—H bonds in each H_2O molecule)

b $CH_4 + 2O_2 \rightarrow CO_2 + 2H_2O$

c

BONDS BROKEN	NUMBER	BONDS FORMED	NUMBER
Between carbon and hydrogen	4	Between carbon and oxygen	2
Between oxygen atoms	2	Between hydrogen and oxygen	4

d Exothermic

3 a $C_3H_8(g) + 10O_2(g) \rightarrow 3CO_2(g) + 4H_2O(g)\ \Delta H = -2220$ kJ mol^{-1}

b $2NH_3(g) \rightarrow N_2(g) + 3H_2(g)\ \Delta H = +92$ kJ mol^{-1}

c $CaCO_3(s) \rightarrow CaO(s) + CO_2(g)\ \Delta H = +180$ kJ mol^{-1}

d $4NH_3(g) + 5O_2(g) \rightarrow 4NO(g) + 6H_2O(g)\ \Delta H = -905$ kJ mol^{-1}

e $H_2O(l) \rightarrow H_2O(s)\ \Delta H = -6.02$ kJ mol^{-1}

4 a $2CO(g) + O_2(g) \rightarrow 2CO_2(g)$

REACTANTS	PRODUCTS
ΔH_f (kJ mol^{-1})	ΔH_f (kJ mol^{-1})
CO −110.53	CO_2 −393.5
O_2	0
Total: −110.53 × 2	Total −393.5 × 2

ΔH(reaction) = Total ΔH_f(products) – Total ΔH_f(reactants) = – 565.94 kJ

Hence $\Delta H\text{(reaction)} = \dfrac{-565.94 \text{ kJ}}{2 \text{ moles}}$ kJ mol^{-1}

b $CH_4(g) + 2O_2(g) \rightarrow CO_2(g) + 2H_2O(l)$

REACTANTS	PRODUCTS
ΔH_f (kJ mol^{-1})	ΔH_f (kJ mol^{-1})
CH_4 – 74.8	CO_2 – 393.5
O_2 0	H_2O(l) – 285.8 × 2
Total: –110.53	Total –965.1

ΔH (reaction) = Total ΔH_f (products) – Total ΔH_f (reactants) = –890.3 kJ mol^{-1}

c $2H_2S(g) + 3O_2(g) \rightarrow 2H_2O(l) + 2SO_2(g)$

REACTANTS	PRODUCTS
ΔH_f (kJ mol^{-1})	ΔH_f (kJ mol^{-1})
H_2S – 20.1 × 2	H_2O(l) –285.8 × 2
O_2 0	SO_2 –296.1 × 2
Total: –40.2	Total –1163.8

ΔH (reaction) = Total ΔH_f (products) – Total ΔH_f (reactants) = – 561.8 kJ mol^{-1}

d $2NO(g) + O_2(g) \rightarrow 2NO_2(g)$

REACTANTS	PRODUCTS
ΔH_f (kJ mol^{-1})	ΔH_f (kJ mol^{-1})
NO + 90.4 × 2	NO + 33.9 × 2
O_2 0	
Total: +180.8	Total + 67.8

ΔH (reaction) = Total ΔH_f (products) – Total ΔH_f (reactants) = –56.5 kJ mol^{-1}

5 a $S(s) + O_2(g) \rightarrow SO_2(g)$ –297 kJ × 2

$2SO_3(g) \rightarrow 2SO_2(g) + O_2(g)$ +198 kJ reverse the reaction

$2S(s) + 2O_2(g) \rightarrow \cancel{2SO_2(g)}$ –594 kJ

$\cancel{2SO_2(g)} + O_2(g) \rightarrow 2SO_3(g)$ –198 kJ

Nett $2S(s) + 2O_2(g) \rightarrow 2SO_3(g)$ –792 kJ

b $Sn + Cl_2 \rightarrow \cancel{SnCl_2}$ –325 kJ

$\cancel{SnCl_2} + Cl_2 \rightarrow SnCl_4$ –186 kJ

Nett $Sn + 2Cl_2 \rightarrow SnCl_4$ –511 kJ

c $O_2 \rightarrow 2O + 495$ kJ $\times \frac{1}{2}$ and reverse

$2O_3 \rightarrow 3O_2 - 427$ kJ $\times \frac{1}{2}$ and reverse

$NO + O_3 \rightarrow NO_2 + O_2$ –199 kJ

$\cancel{O} \rightarrow \cancel{\frac{1}{2}}\cancel{O_2}$ 247.5 kJ

$\cancel{\frac{3}{2}}\cancel{O_2} \rightarrow \cancel{O_3}$ + 213.5 kJ

$NO + \cancel{O_3} \rightarrow NO_2 + \cancel{O_2}$ –199 kJ

Nett $NO + O \rightarrow NO_2$ –233 kJ

d $4Al + 3O_2 \rightarrow 2Al_2O_3$ –3352 kJ

$Mn + O_2 \rightarrow MnO_2$ –521 kJ × 3 and reverse

$4Al + \cancel{3O_2} \rightarrow Al_2O_3$ –3352 kJ

$3MnO_2 \rightarrow Mn + \cancel{3O_2}$ +1563 kJ

Nett $4\,Al + 3MnO_2 \rightarrow 2Al_2O_3 + 3Mn$ –1789 kJ

6 a

FUEL (1G)	TEMPERATURE AT START (°C)	TEMPERATURE AT END (°C)	TEMPERATURE RISE (°C)
Methanol	23	54	31
Ethanol	19	54	35
Propanol	26	63	37
Butanol	21	61	40

Butanol releases the most energy per gram.

b

FUEL	MASS OF BURNER AT START (g)	MASS OF BURNER AT END (g)	MASS OF FUEL USED (g)
Methanol	145.5	141.9	3.6
Ethanol	202.1	199.3	2.8
Propanol	177.5	175.1	2.4
Butanol	226.3	224.1	2.2

Since the same amount of energy is required to heat 200 g of water by 10°C, the amount of energy per gram is clearly greatest for butanol since it required the least mass to produce the same quantity of heat.

7 Heat absorbed by water = $Q = mC_w\Delta T$

$= 98.5\,(4.186)(30.3 - 21.0)$

$= 3835$ J

Hence heat released by metal sample 3835 J

Mass of sample = 250 g

Rearrange $Q = mC_{metal}\,\Delta T$

$$C_{metal} = \frac{Q}{m\Delta T} = \frac{3835}{250\,(96.0 - 30.3)} = 0.233\ \text{J g}^{-1}\,{}^{\circ}\text{C}^{-1}$$

Hence, the metal was cadmium (given specific heat = 0.232 J/g°C)

9780170412391

EVALUATION

1 a–e

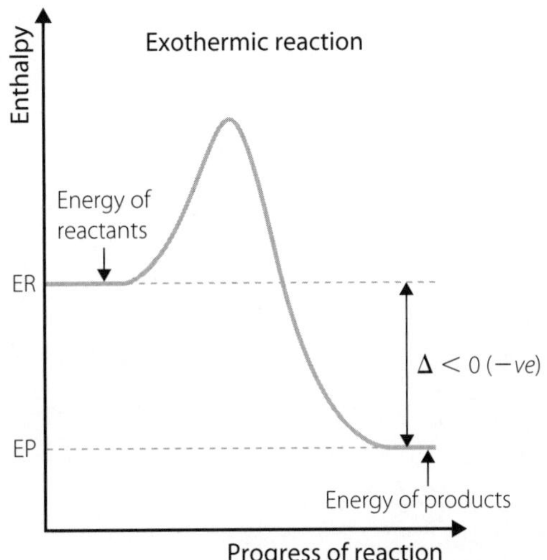

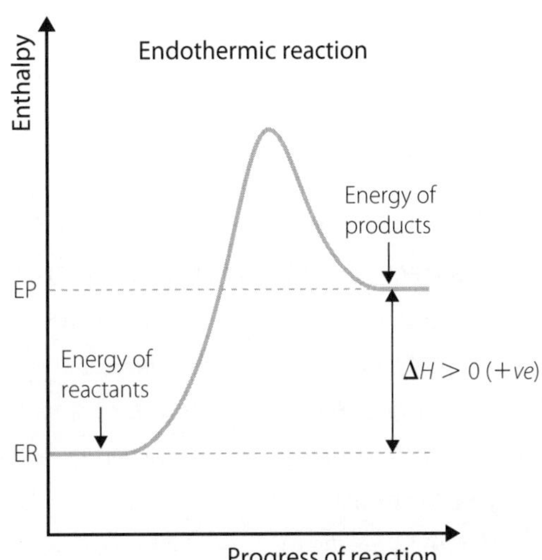

f The 'hump' in each graph is called the activation energy, Ea. It is the energy that needs to be overcome for the reaction to proceed. Activation energy can be defined as the minimum energy required by the reactants to start the chemical reaction to form the products.

g The exothermic reaction shown would not occur spontaneously since there is an energy barrier (the activation energy, Ea) which needs to be overcome before the reaction can proceed.

h The graph for a spontaneous exothermic reaction would have no activation barrier.

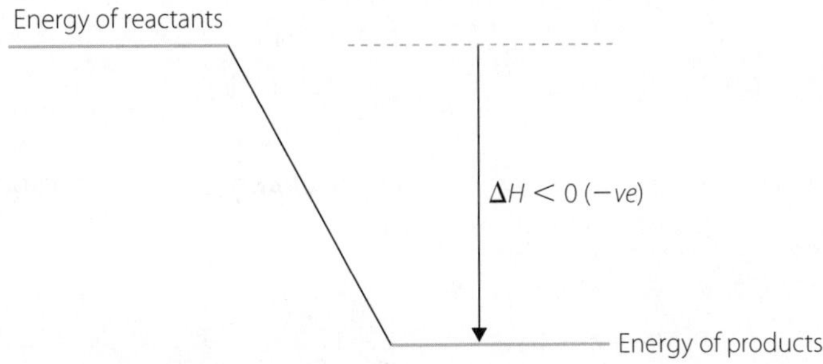

2 Using Hess's Law: $\Delta H = +15.3$ kJ

3 a 837 J

b 837 J

c 11.8 J K^{-1}

d 0.39 J g^{-1} K^{-1}

e Cu (0.38 and 0.39 are consistent within experimental uncertainty, i.e. ± 0.05°C)

CHAPTER 10 MEASUREMENT UNCERTAINTY AND ERROR

10.1 IMPORTANT TERMS

Across

4 Calibration

5 Reproducible

6 Independent

8 Resolution

10 Systematic

12 Precision

14 Dependent

15 Random

Down

1 Uncertainty

2 Measurement

3 Outlier

7 True value

9 Regression

10 Standard

11 Accuracy

13 Average

10.2 MAKING MEASUREMENTS, PRECISION VERSUS ACCURACY

1 a Precision: 1 minute (5%) Uncertainty: ± 1 minute (± 5%)

b Precision: 0.1°C (0.4%) Uncertainty: ± 0.05°C (± 0.2%)

c Precision: 0.1°C (0.3%) Uncertainty: ± 0.1°C (± 0.3%)

d Precision: 0.001 g (0.07%) Uncertainty: ± 0.001 g (± 0.07%)

e Precision: 1 m (0.8%) Uncertainty: ± 0.5 m (± 0.4%)

f Precision: 0.001 nm (0.8%) Uncertainty: ± 0.0005 nm (0.4%)

2 a 0.01 s (0.02%)

b ± 0.01 s (± 0.02%)

c 56.64 ± 0.01 s

d Maximum – 56.65 s, minimum – 56.63 s

3

MEASUREMENT	PRECISION (%)	√	MEASUREMENT	PRECISION (%)	√
(1 ± 0.001) nm	0.1		1 kilometre ± 0.1 metre	0.01	√
7 hours ± 15 minutes	3.6		1 year ± 1 day	0.27	
(25 ± 0.1) ml	0.4		(1000 ± 1) ml	0.1	
(235.6 ± 0.1) kg	0.04	√	1.364 ± 0.005 g	0.37	

10.3 QUALITATIVE AND QUANTITATIVE DATA

DATA	QUALITATIVE OR QUANTITATIVE?
500 ml of ethanol contained in a 1L beaker	Quantitative
Describing construction material for a beaker (e.g. glass, plastic)	Qualitative
Classifying room air (e.g. dry, humid, warm, cold)	Qualitative
Stating the mass of a weighed sample as 5.10 g	Quantitative
The colour of a rock sample	Qualitative
The surface area per gram of a titanium(IV) oxide (TiO_2) nanosphere = 179.9 $m^2\ g^{-1}$	Quantitative
A rise of 21.2°C in the temperature of a calorimeter	Quantitative
A molecular weight of 35.5 g	Quantitative
The three-dimensional geometric shape of molecules (e.g. octahedral, linear or trigonal planar)	Qualitative
The wavelength of blue light from an argon laser quoted as 488 nm	Quantitative

10.4 RANDOM AND SYSTEMATIC ERRORS

1

MASS BEING MEASURED	BALANCE 1 (g)	BALANCE 2 (g)	AVERAGE MASS WITH UNCERTAINTY
Empty beaker	223.92 ± 0.01	227.15 ± 0.01	225.54 ± 0.02
Beaker + 100 ml KCl solution	325.92 ± 0.01	327.15 ± 0.01	326.54 ± 0.02
100 ml KCl solution (by difference)	102.00 ± 0.02	100.00 ± 0.02	101.00 ± 0.04

c Systematic (balance accuracies are undefined).

d Calibration of each balance is required.

2

	TRIAL 1 (g)	TRIAL 2 (g)	AVERAGE MASS WITH UNCERTAINTY
Empty beaker	223.9 ± 0.1	224.4 ± 0.1	224.2 ± 0.2
Beaker + 100 ml KCl solution	323.6 ± 0.1	324.9 ±0.1	324.3 ± 0.2
100 ml KCl solution (by difference)	99.7 ± 0.2	100.5 ±0.2	100.1 ± 0.4

c Random (trials give demonstrably different results).

d A measuring cylinder was used to deliver the sample. Estimating a volume of 100 mL with a measuring cylinder would have uncertainty because of the sighting of the meniscus as well as uncertainty in the delivery. A measuring cylinder (wide cross-section) is not designed like a pipette, which delivers a more precise volume because of its narrow capillaries at each end: at the top for estimating volume, and at the bottom to deliver the substance.

e It is impossible to determine which of the two trials provides the more reliable measurement. Repeated trials would yield an improved estimate of the uncertainty via a calculation of the standard deviation.

10.4 SIGNIFICANT FIGURES, ABSOLUTE UNCERTAINTY, PERCENTAGE UNCERTAINTY, PERCENTAGE ERROR

1 a 4.0101×10^7

b 8.912×10^{-3}

c Correct

2 a i 55.85 mL ± 0.04%

ii 6.78 cm ± 0.3%

b Convert the following percentage uncertainties to absolute uncertainties:

i 1.45 L ± 1% = 1.45 ± 0.01 L

ii 32.33 mm ± 0.2% = 32.33 ± 0.05 mm

c i (2.95 ± 0.01) L + (0.563 ± 0.005) L = 3.51 ± 0.02 L

ii (99.99 ± 0.02) g − (17.45 ± 0.01) g = 82.54 ± 0.03 g

iii (2.95 ± 0.01) L × (0.563 ± 0.005) L = 1.66 L ± 1.2%

iv (99.99 ± 0.02) g ÷ (17.45 ± 0.01) g = 5.73 g ± 0.1%

v (2.95 ± 1%) L + (0.563 ± 0.2%) L = 3.51 L ± 1.2%

vi (99.99 ± 0.5%) g − (17.45 ± 0.05%) g = 82.54 L ± 0.6%

vii (2.95 ± 1%) L × (0.563 ± 0.2%) L = 1.66 L ± 1.2%

viii (99.99 ± 0.5%) g ÷ (17.45 ± 0.05%) g = 5.73 g ± 0.6%

ix 2 × (2.95 ± 0.01) cm = 5.90 L ± 0.7%

x 2 × (33 s ± 2%) cm = 66 s ± 4%

xi $(2.95 \pm 0.01\ \text{cm})^2 = 8.9\ \text{cm}^2 \pm 0.7\%$

xii $\frac{4}{3}\pi R^3$ where $R = 6{,}371 \pm 1$ km $= 1.0832 \times 10^{12}$ km ± 0.05%

The statement is not appropriate, as the range between mountains and oceans is up to 8.8 km (Everest) and, furthermore, the equatorial radius is approximately 22 km greater than the polar radius. The radius should be quoted as 6370 ± 20 km.

10.6 PRESENTING AND ANALYSING DATA

1 Temperature (°C)

2 Specific heat capacity ($\text{J K}^{-1}\,\text{g}^{-1}$)

3 Domain: $0 \le T \le 100$°C; Range: 4.175 to 4.220

4 (Plot on graph paper)

5 (and **7**) – see graph below

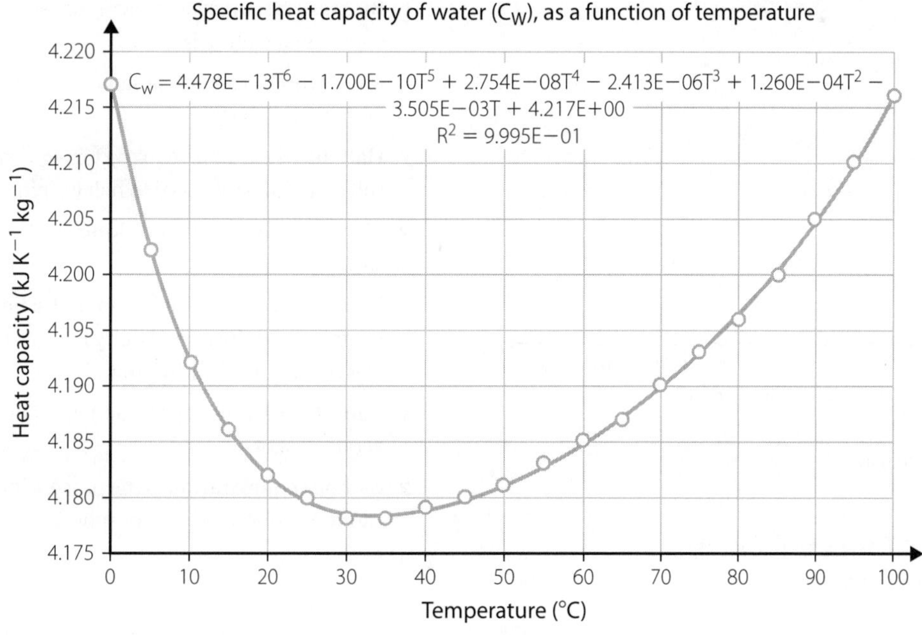

6 From the graph, the specific heat capacity of water is found to be a minimum (4.178 $\text{J K}^{-1}\text{g}^{-1}$) at 31.3°C

8 Below 0°C water freezes and above 100°C water vaporizes and the specific heat capacity changes significantly. For example, the specific heat capacity for water vapour at 100°C is 2.080 (see table).

9 Based on either a 4th or 6th order polynomial model, C_w at 23°C is determined to be 4.180 $\text{J K}^{-1}\text{g}^{-1}$.

10 The uncertainty of ±0.0005 $\text{J K}^{-1}\text{g}^{-1}$ for C_w converts to ±0.012%. This is combined with the greater uncertainty of at best ±0.5 K for temperature measurements (i.e. 0.13% – 0.18% relative to T = 373K and 273 K respectively). Hence, we calculate the overall uncertainty as: ± (0.18 + 0.012)% = ± 0.2%

11 Mean = μ = 4.191 J $K^{-1}g^{-1}$; standard deviation = σ = ± 0.013 J $K^{-1}g^{-1}$. The normally quoted value of 4.184 J $K^{-1}g^{-1}$ lies within the range of $\mu \pm \sigma$, i.e. 4.171 – 4.197 J $K^{-1}g^{-1}$. The value of 4.184 J $K^{-1}g^{-1}$ normally quoted is a better estimate for C_w in the lower range of typical air temperatures, i.e. 15–20°C.

EVALUATION

1 **a** 4

b 5

c 3

d 3

2 **a** 104.3 g

b 57.0

3 Final concentration: 0.00753 M ± (% uncertainty) or 0.00753 ± 0.00006 M (absolute uncertainty)

4 M = 33.981 ± 0.004 g (or M = 33.981 g ± 0.003%)

5 Heat absorbed = 2200 ± 220 J (± 10%)

6 Mean (μ) = 0.1006 moles L^{-1}

Standard deviation (σ) = 0.0010

Relative standard deviation $\left(\frac{\sigma}{\mu}\right) \times 100\% = 1\%$ (rounded from 0.98%)

7 **a, b**

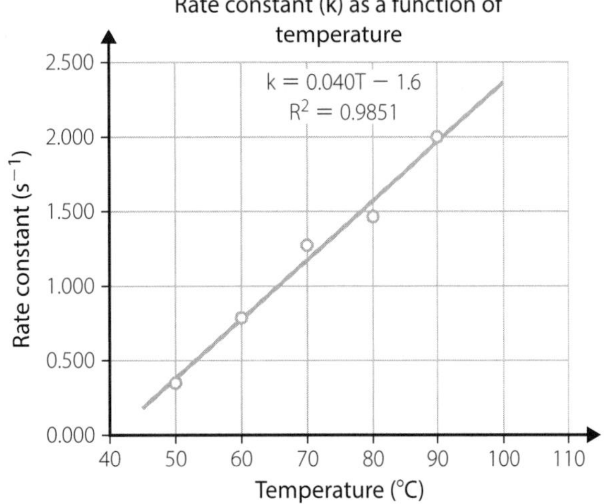

c Gradient = 0.040

d ± 0.005

e K = (0.040 ± 0.005) T – 1.6 (± 0.05)

f At T = 100, k = 2.40 ± 0.01

CHAPTER 11 FUELS

11.1 IMPORTANT TERMS

1 **a** The **density** of a substance is its mass divided by its volume.

b The **viscosity** of a liquid is its resistance to flow.

c The **volatility** is the tendency of a substance to vaporise.

d **Vapour** is the gaseous phase of water.

e **Alpha radiation** is emission of a helium nucleus.

f **Gamma radiation** is short wavelength, high-energy radiation that will pass through flesh and bone.

g **Infrared radiation** is invisible to the human eye, emitted by all objects and sensed as heat.

h A **fuel** is burnt to produce heat or **power**.

i There are three major sectors that use fuels: **domestic**, **industrial** and **transport**.

j Fossil fuels are produced from **organic** matter that has **decomposed** over a long time under high pressure and temperature.

k Sulfur is an **impurity** found in fossil fuels.

l No fuel has 100% **efficiency**.

m A **plasma** contains a sea of charged particles because electrons are free to move independently of the nucleus.

n A **hydrocarbon** compound contains only carbon and hydrogen atoms.

o Nuclear fusion is joining of small **nuclei**

p The optimal ratio of energy outputs to energy inputs for a nuclear fusion power plant is known as its **performance**

q A **toroidal** field is generated by wrapping coils around the poles of a sphere or donut shape.

r A **porodial** field is generated by wrapping coils around the lines of latitude of a sphere or donut shape.

s There are two types of **smog**: traditional (London) and photochemical (Los Angeles).

t **Biomass** is sourced from living or recently living plant and animal materials.

u A **neutron** has about the same mass as a proton, but without any charge.

v **Half–life** is the time required for a quantity of a substance to be reduced to half its initial value.

w Thomas **Edison** generated **kinetic** energy to drive a steam train.

x Michael **Faraday** used the **magnetic field** of a magnet to produce **electricity** in a coil of wire via **electromagnetic induction**.

y **Photolysis** is a chemical reaction occurring by the action of light.

z Carbon **sequestration** is the process involved in the capture and storage of atmospheric carbon dioxide.

2 **a** H

b C

c A

d E

e D

f B

g F

h I

i G

3

FUEL	EQUATION	ADVANTAGES	DISADVANTAGES
Coal	$C(s) + O_2(g) \rightarrow CO_2(g)$	Tried and tested	Large CO_2 emissions, impurities
Petrol	$2C_8H_{18}(l) + 25O_2(g) \rightarrow 16CO_2(g) + 18H_2O(l)$	Less sulfur and nitrogen oxides and dioxides	More unburnt hydrocarbons
Diesel	$2C_{16}H_{34} + 49O_2 \rightarrow 32CO_2 + 34H_2O$	Less unburnt hydrocarbons	More sulfur and nitrogen oxides and dioxides
Natural gas	$CH_4 + 2O_2 \rightarrow CO_2 + 2H_2O$	Less CO_2 emissions, fewer impurities	Potential explosions of wells and pipelines
Hydrogen	$2H_2 + O_2 \rightarrow 2H_2O$	Only product is water	Low density of hydrogen
Fission	${}^{1}_{0}n + {}^{235}_{92}U \rightarrow {}^{137}_{52}Te + {}^{97}_{40}Zr + 2{}^{1}_{0}n$	Cheap, no toxic emissions	Extraction of fuel, storage of waste, reactor explosions
Fusion	${}^{2}_{1}H + {}^{3}_{1}H \rightarrow {}^{4}_{2}He + {}^{1}_{0}n$	No toxic emissions	Currently low performance
Biofuel	$C_2H_5OH + 3O_2 \rightarrow 2CO_2 + 3H_2O$	Renewable	Large CO_2 emissions, low efficiency

4

Across

1 biofuel
3 hydrogen
5 fission
6 uranium
7 lithium
8 sodium
10 nuclear
12 nitrogen
14 fusion
18 renewable
19 sulfur
20 diesel
21 methanol

Down

2 fossil fuels
3 helium
4 ethanol
9 non-renewable
11 reactor
13 electricity
15 natural gas
16 combustion
17 greenhouse

11.2 EARTH-LIKE PLANET SURFACE TEMPERATURES

1, 2 Students' own responses

11.3 SNOWFALL ACIDIFICATION

1–3 Students' own responses

EVALUATION

1 C
2 D
3 B
4 D
5 non-renewable
6 Ethanol
7 Leakage
8 Lack of technology
9 Coal and oil contain impurities such as sulfur. When these are burnt as part of the coal and oil, they produce gases which dissolve in water vapour to give acid rain

$S(s) + O_2(g) \rightarrow SO_2(g)$

$SO_2(g) + H_2O(g) \rightarrow H_2SO_3(aq)$

10 Hydrogen is produced from methane by reacting the methane with steam:

$CH_4(g) + 2H_2O(g) \rightarrow 4H_2(g) + CO_2(g)$

Possible environmental problem is the production of CO_2 gas

11 a Methanol: $2CH_3OH + 3O_2 \rightarrow 2CO_2 + 4H_2O$

Methane: $CH_4 + 2O_2 \rightarrow CO_2 + 2H_2O$

Ethanol: $C_2H_5OH + 3O_2 \rightarrow 2CO_2 + 3H_2O$

Ethane: $2C_2H_6 + 7O_2 \rightarrow 4CO_2 + 6H_2O$

Butanol: $2C_4H_9OH + 13O_2 \rightarrow 8CO_2 + 10H_2O$

Petrol: $2C_8H_{18} + 25O_2 \rightarrow 16CO_2 + 18H_2O$

The energy density and CO_2 values for each fuel for must be quoted.

General trends must be given:

The alcohols have a lower energy density than their matching alkane.

As the number of carbons increase for both alkane and alcohols, the energy density increases. However, the CO_2 also increases as number of carbon atoms increase.

b Alcohols currently are not viable. However, since energy production increases with increasing number of alcohols, if a method could be developed to produce longer chain alcohols cheaply, then biofuels would be viable.

CHAPTER 12 MOLE CONCEPT AND THE LAW OF CONSERVATION OF MASS

12.1 IMPORTANT TERMS

1 a Anions – **L**
  b Atomic mass unit – **S**
  c Species – **D**
  d Avogadro constant (N_A) – **N**
  e Balanced equation – **O**
  f Cations – **Q**
  g Chemical equation – **Y**
  h Chemical formula – **F**
  i Empirical formula – **V**
  j Formula unit – **E**
  k Ionic compound – **B**
  l Lattice – **T**
  m Mass (m) – **R**
  n Molar mass (M) – **J**
  o Mole of a substance – **A**
  p Molecular compound – **C**
  q Molecular formula – **G**
  r Molecule – **H**
  s Percentage composition – **P**
  t Products – **M**
  u Reactants – **W**
  v Relative atomic mass (A_r) – **I**
  w Relative formula mass (M_r) – **U**
  x Relative molecular mass (M_r) – **K**
  y Stoichiometry – **X**

12.2 MOLE AND AVOGADRO'S NUMBER OF PARTICLES

1

SUBSTANCE	MOLAR MASS (g) (M)	NUMBER OF PARTICLES ($n \times N_A$)	NUMBER OF MOLES OF PARTICLES (n)	MASS OF n MOLES (g)
Germanium (Ge)	72.6	1.51×10^{24}	2.5 Moles of germanium atoms	182
Methanoic (formic) acid (HCOOH)	46.0	1.81×10^{24}	3.0	138 g of methanoic acid
Calcium phosphate ($Ca_3(PO_4)_2$)	310.2	2.409×10^{24} molecules of calcium phosphate	4.0	1241
Xenon hexafluoride (XeF_6)	245.3	1.81×10^{24}	3.0 Moles of xenon hexafluoride molecules	736
Dichloromethane (CH_2Cl_2)	84.9	1.81×10^{24}	3.0	255 g of dichloromethane

2 6.022×10^{23}
3 6.022×10^{23}
4 35.45 g
5 6.022×10^{23}
6 1.2044×10^{24}
7 9.04×10^{22}
8 71 g
9 18.0 g
10 294.2 g
11 342.2 g
12 a 2
  b 2
  c 6
  d 3.61×10^{24}

12.3 EMPIRICAL AND MOLECULAR FORMULAS

1 $Na_2Cr_2O_7$
2 Cr_2O_3
3 41.16% N, 47.00% O, 11.84% H
4 C_6H_{12}
5 C_6H_5MgBr; empirical and molecular formulas are the same.

12.4 PERCENTAGE YIELD AND THEORETICAL YIELD

1 a Not balanced: $C_2H_5OH(l) + 3O_2(g) \rightarrow 2CO_2(g) + 3H_2O(l)$
  b O_2 (Since 46 g C_2H_5OH is equivalent to 1 mole, 64 g O_2 is equivalent to 2 moles but 3 moles of O_2 are required to react with 1 mole C_2H_5OH to satisfy the reaction stoichiometry. Hence C_2H_5OH is in excess and only $\frac{2}{3}$ moles = 30.7 g of the 46 g will be used up in the reaction when 64 g of O_2 has been consumed).

9780170412391

c 58.7 g if only 64 g O_2 is available and C_2H_5OH is in excess in the reaction given.

2 a 34.4 g LiCl

b Percentage yield = $\frac{6}{34.4}(100)\% = 16.9\%$

3 a $H_3PO_4 + 3KOH \rightarrow K_3PO_4 + 3H_2O$

b 108.2 g K_3PO_4

c 63 g K_3PO_4 corresponds to 58.3% yield

EVALUATION

1 a False – the relative atomic mass of an element is defined as the ratio of the weighted average mass per atom of the naturally occurring form of the element to 1/12 the mass of an atom of carbon-12.

b True

c False – All pure substances exist as atoms or molecules

d True

e True

f False – one mole of a substance would contain $6.022 \times 10^{+23}$ particles of that substance. (Note the index is +23, not −23!))

g True

h False – molar mass is an actual mass that is measured in grams.

i True

j False – stoichiometry focuses on the molar ratios of reactants and products in a reaction.

2 a 17.04

b 98.08

c 62.03

3 a 142.04

b 148.33

c 58.33

4 0.23 g NO_2

5 a $K_2PtCl_4 + 2NH_3 \rightarrow Pt(NH_3)_2Cl_2 + 2KCl$

b 245 g KCl

c NH_3

d 0.51% is the experimental percentage yield of $Pt(NH_3)_2Cl_2$ if 2.5 g of $Pt(NH_3)_2Cl_2$ are collected after starting with 56 g of NH_3 and an excess of K_2PtCl_4.

CHAPTER 13 MOLECULAR INTERACTIONS AND REACTIONS

13.1 IMPORTANT TERMS

1

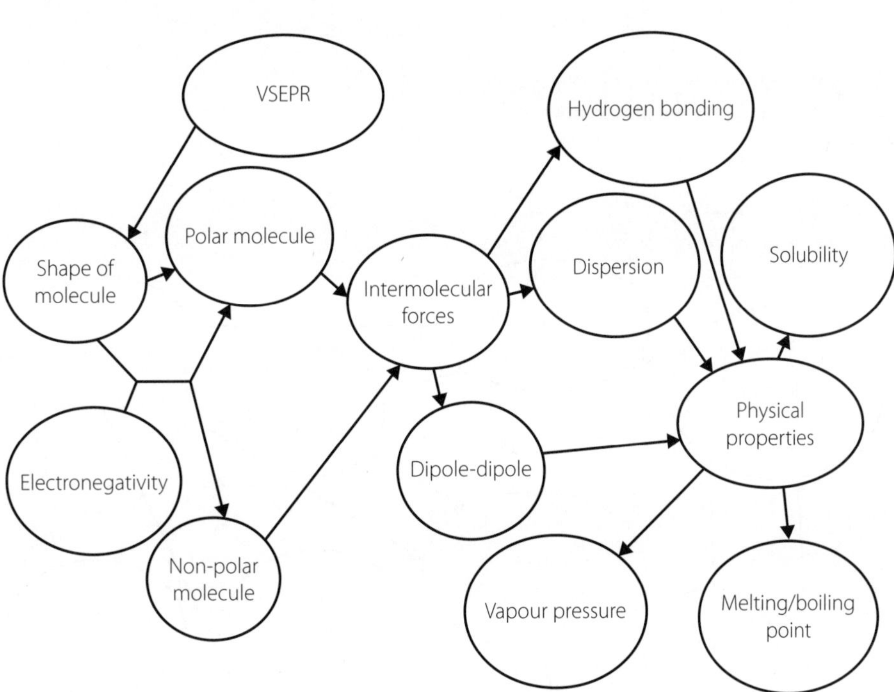

13.2 VALENCE SHELL ELECTRON PAIR REPULSION (VSEPR) THEORY

1 a H_2S = bent (or V – shaped)

b BH_3 = trigonal planar

c HCl = linear

d SiH_4 = tetrahedral

e SO_2 = bent

f NO_2 = bent

2 a Water (H_2O) has a **tetrahedral** arrangement of electron pairs around the central O atom but its actual shape is **bent** because of the presence of **lone pairs** around the O atom.

b Ammonia (NH_3) has a **tetrahedral** arrangement of electron pairs around the central N atom, whereas boron trihydride (BH_3) has a **trigonal planar** arrangement of electron pairs around the central B atom. This explains why NH_3 is **trigonal pyramidal** shaped and BH_3 is **trigonal planar** shaped.

c If a molecule had a general formula of MX_3 it would be **trigonal planar** shaped but if it had the formula MX_2E it would be **bent**, even though it contained the same number of electron pairs around the central atom. Likewise, a tetrahedral arrangement could be produced by a MX_4, a $\mathbf{MX_3E}$, or a $\mathbf{MX_2E_2}$.

13.3 POLARITY OF MOLECULES

1 P – F(1.8), C – F(1.5), O – H(1.4), S – F(1.3), C – O(1.0), H – Cl(0.8), N – H(0.8), C – Br(0.3), F – F(0.0)

EVALUATION

1 D

2 C

3 B

4 dispersion

5 atomic mass

6 CH_2O contains hydrogen bonding, BF_3 is trigonal planar (some explanation needs to be given for this) and is non-polar due to its molecular geometry (the B – F bond is polar, the dipole being 2.0). Therefore BF_3 contains dispersion forces only and so very little heat energy needs to be supplied to boil it.

7 SO_2 exhibits dipole – dipole interactions (bent), CO_2 exhibits dispersion forces only linear). SO_2 has stronger intermolecular forces than CO_2 and so will have a higher boiling point and vapour pressure. Due to the principle of like dissolves like, SO_2 (polar substance) will dissolve better in water (polar substance) than CO_2 (non-polar substance).

8 a–c

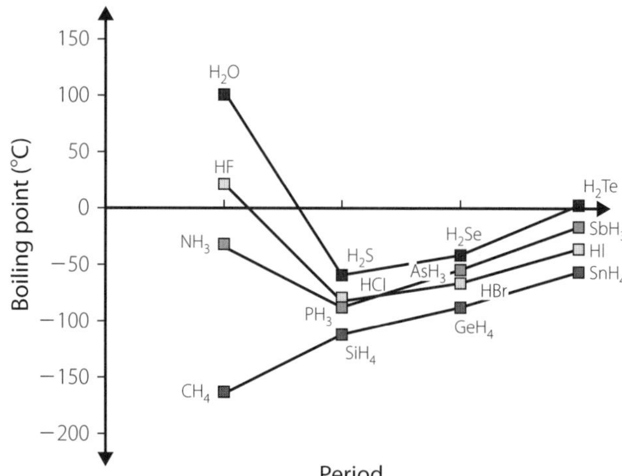

b Answers should describe upward trend from periods 2 – 5 (increasing molecular size). Group 7 higher than group 4 (polarity) but lower than Group 6 (greater polarity). Groups 6 and 7 have very high boiling period. 2 hydrides due to hydrogen bonding.

CHAPTER 14 CHROMATOGRAPHY TECHNIQUES

14.1 IMPORTANT TERMS

1

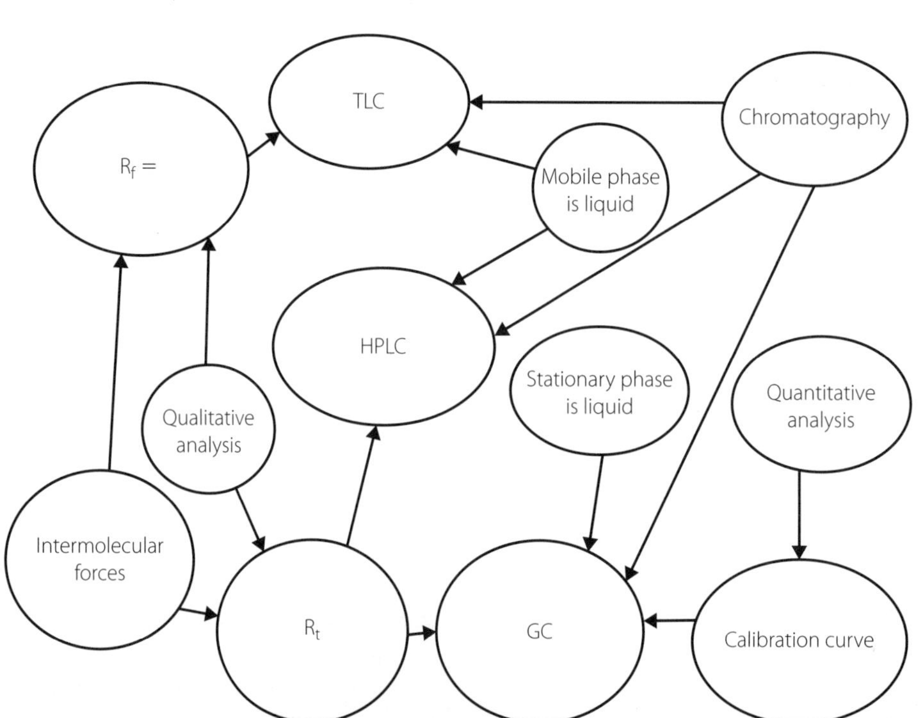

9780170412391

14.2 CHROMATOGRAPHY TECHNIQUES

1 a A solid or a solid coated in a liquid polymer.

b The ratio of the distance travelled by an analyte from the origin to the distance travelled by the mobile phase.

c The release from a substance from the surface of a solid.

d The time taken for an analyte to travel through a GC or HPLC column to a detector.

e A method of determining the quantity of a substance in an unknown sample by measuring an instruments' (GC or HPLC) response known amounts of the substance, plotting a graph of these responses and comparing the quantity to it.

f The attraction of a substance to the surface of a solid.

2 a In TLC, the **stationary** phase is usually a fine powder of silica attached to glass or plastic while the **mobile** phase is a solvent in which the **analyte** is dissolved.

b Separation of substances depends on the **component** present in **intermolecular forces** and how strongly they **absorb** onto the surface of the plate.

c In **gas chromatography** the **mobile** phase is an inert gas. This method is better suited to substances that can be easily **vaporised** unlike **HPLC** which is more appropriate for molecules that are unstable at high temperatures. This method requires the use of **pumps** to move the **mobile** phase through the column.

14.3 SEPARATING COMPONENTS

1

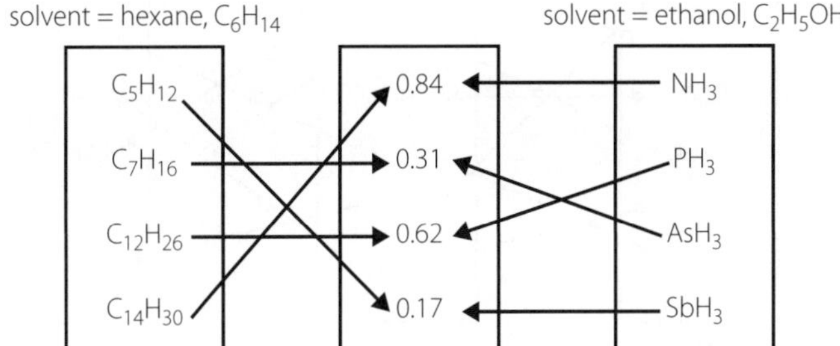

EVALUATION

1 A

2 B

3 Desorption

4 Polar

5 The analytes would only be very slightly, if at all, solvent in a non-polar solvent and so will have very similar R_f values.

6 A calibration graph should be used to compare the gas chromatography response (peak area) to the unknown quantity of the component with a series of known amounts of the same component.

7 HPLC is a better method because GC would probably destroy the component and TLC would only be able to give qualitative results.

8 The distance each component has moved from the origin needs to be determined using the R_f equation:

1 = 1.0 cm

2 = 2.9 cm

3 = 3.1 cm

4 = 5.8 cm

Using these distances and the distance of the solvent front, the spots and the solvent front need to be drawn clearly on the diagram of a TLC plate provided.

CHAPTER 15 GASES

15.1 IMPORTANT TERMS

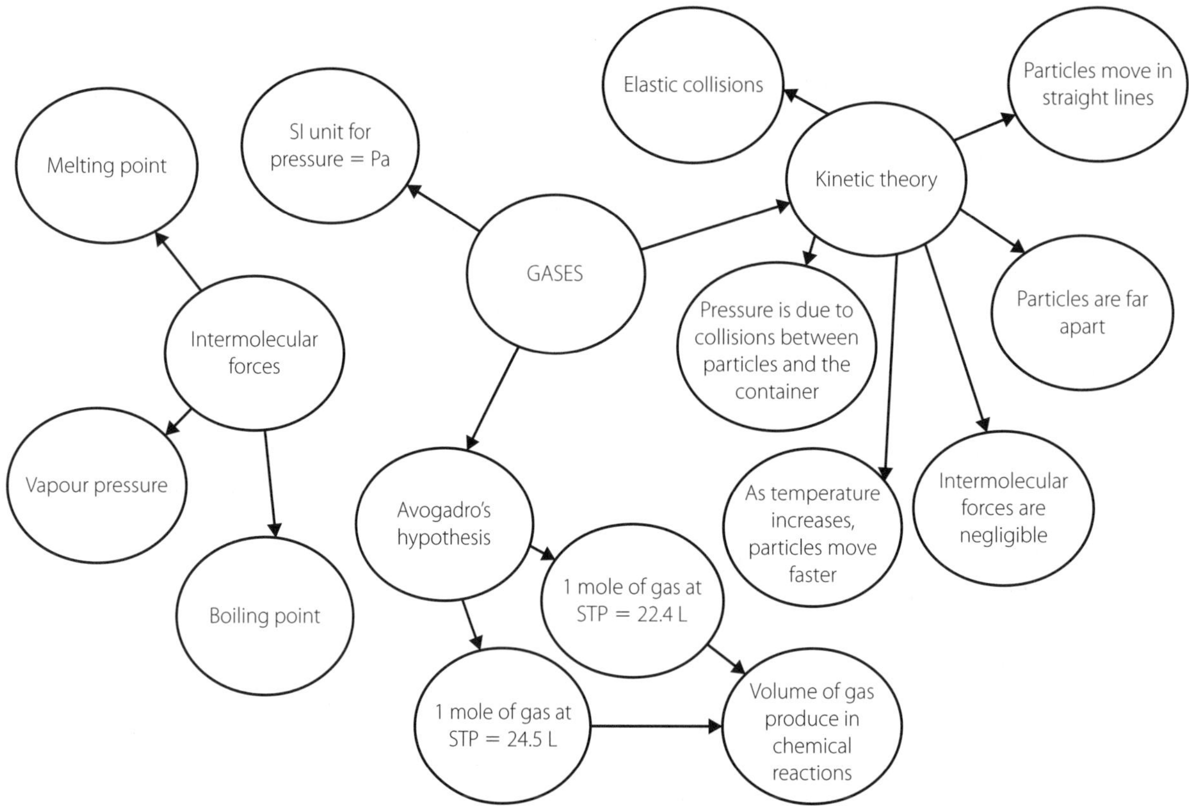

15.2 GASES

1 a The pressure exerted by a vapour in equilibrium with its solid or liquid phase.

b The volume occupied by one mole of a gas at standard temperature and pressure.

c The values of 0°C and 1 atmosphere.

d Equal volumes of gas under the same conditions of temperature and pressure contain the same number of particles.

e SI unit of pressure equal to a force of 1 Newton per square metre.

f The values 25°C and 1 atmosphere.

g A gas that conforms to the kinetic theory of gases.

2 a **Avogadro's** hypothesis states that equal **volumes** of gases contain **equal** numbers of particles.

b One mole of any gas occupies **22.4** L at STP, which is **0**°C and **1** atm pressure.

c At SLC the temperature and pressure is **25**°C and **1** atm respectively.

15.3 KINETIC THEORY

1 a noble gases

b large

c negligible

d elastic energy

e container

2

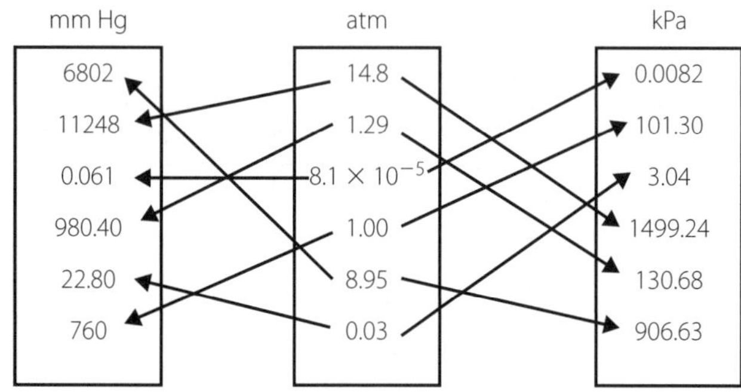

EVALUATION

1 C

2 D

3 Molecular mass

4 The particles of gas become cooler and so have less kinetic energy. The force with which they collide with the interior of the balloon is less and so the balloon contracts.

9780170412391

5 a

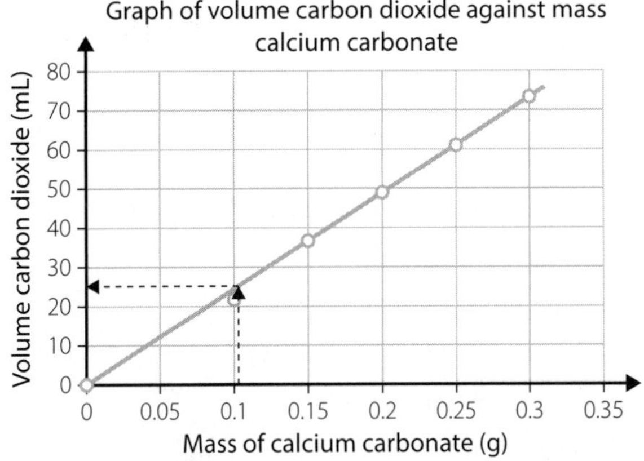

b 0.1 g = n($CaCO_3$)/1000 = 25 mL = molar volume/1000

From the balanced equation, 1 mole $CaCO_3$ = 1 mole CO_2

Therefore, molar volume = 25 L at SL

c 2.04%

d Some of the CO_2 may have dissolved in the water; measurement errors such as mass balance, reading errors or volume reading errors.

CHAPTER 16 AQUEOUS SOLUTIONS AND MOLARITY

16.1 IMPORTANT TERMS

Across

4 Vapour
6 Solute
7 Capacity
9 Hydrophilic
10 Molarity
11 Density
12 Hydrophobic

Down

1 Capillary
2 Vaporisation
3 Solubility
5 Tension
7 Concentration
8 Anhydrous

16.2 WATER PROPERTIES AND STRUCTURE

1 Ethane is non-polar. The structure is symmetric with a completely symmetric electron distribution, hence no net build-up of excess negative or positive charge anywhere in the molecule.

2 Ethanol has high electron density on the oxygen atom because of the two lone pairs on the O atom. Thus, those lone pairs can form hydrogen bonds with the H atoms on other ethanol molecules or the H atoms on water. Likewise, the —OH group has its H atom influenced by the electronegative O atom, so is 'delta positive' (δ+) and can form hydrogen bonds with O atoms in water molecules or in other ethanol molecules.

3 A hydrogen bond is an attraction between polar molecules in which hydrogen is bound to a highly electronegative atom. This causes strong dipole-dipole interactions. The strong bonding with hydrogen in another molecule only happens with the elements N, O and F. The hydrogen bonds are strongest when the H atoms involved are themselves bonded to the highly electronegative atoms N, O and F. In the case of water (H_2O), the hydrogen bonding between water molecules in the liquid phase is very strong, hence liquid water consists of a network of strongly hydrogen bonded water molecules.

Each water molecule can potentially form four hydrogen bonds with surrounding water molecules, hence that leads to a small 'polymer' of five tightly bonded water molecules. As liquid water is really a continuously varying polymeric structure, it has, on average, a much higher effective molecular weight than one water molecule. Accordingly, it takes much more kinetic energy (heat) for individual water molecules to be separated from their networks and released into the gas phase, or boil. If we contrast H_2O with H_2S, the latter has a molecular weight of 34 compared to water's 18, but since H_2S does not hydrogen bond to any significant extent, it has a 'normal' boiling point of –60°C, which is consistent with its molecular weight. (The electronegativity of S is 2.5, O is 3.5).

4 covalent bonds between the central oxygen atom and the two hydrogen atoms

5 polar covalent bonds; in water, these bonds are among the strongest polar covalent bonds in nature and are given the specific name 'hydrogen bonds'

6 the intermolecular forces in liquid water of which the strongest are the hydrogen bonds between water molecules

16.3 SOLUTES, SOLVENTS, SOLUTION AND CONCENTRATION

1 Solute is the ionic salt sodium chloride (NaCl); the solvent is water (H_2O).

2 a 0.10 molal (i.e. moles/100g solvent)

b 74.6 g/L (taking the density of water to be 1.00 g cm^{-3}) hence 7.46 W/V%

c 1.0 Molar (moles L^{-1})

3 a 0.003 W/V%

b 0.00017 M

EVALUATION

1 C
2 D
3 A
4 D

5 **a** Concentration of saline is 0.09% m/v which means 0.09 g NaCl per 100 mL of water. Hence a capsule containing just 5 mL of the saline solution will contain:

$$\frac{5}{100} \times 0.09 = 0.0045\,\text{g} \text{ or } 4.5\,\text{mg}$$

b When 495 mL water is added to 5.0 mL of 8.0 $molL^{-1}$ sodium chloride solution, the new total volume is 500 mL. Hence the sodium chloride solution will be diluted by the *factor*: $\frac{5}{100} = \frac{1}{100}$ Note that the number of moles of NaCl that was originally in the 5.0 mL of solution has not changed. Only the total volume which contains that number of moles has changed.

So the concentration of the sodium chloride solution will now be:

$$\frac{1}{100} \times 8.0 = 0.08\,\text{molL}^{-1}$$

CHAPTER 17 IDENTIFYING IONS IN SOLUTION

17.1 IMPORTANT TERMS

Across

5 anion
6 spectator
9 insoluble

Down

1 overall
2 ion
3 cation
4 ionic
7 precipitate
8 suspension

17.2 USING THE SOLUBILITY RULES

1 **a** soluble
b insoluble
c soluble
d soluble
e soluble
f insoluble
g insoluble
h soluble
i insoluble
j soluble
k insoluble
l soluble
m soluble
n insoluble
o insoluble
p soluble
q insoluble
r soluble
s soluble
t soluble

17.3 PREDICTING AND WRITING CHEMICAL REACTIONS

1 **a** A reaction occurs

Ionic: $Ag^+(aq) + Cl^-(aq) \rightarrow AgCl(s)$

Balanced: $AgNO_3(aq) + NaCl(aq) \rightarrow AgCl(s) + NaNo_3(aq)$

Silver chloride precipitate

b No reaction

c A reaction occurs

Ionic: $Ag^+(aq) + Br^-(aq) \rightarrow AgBr(s)$

Balanced: $Ch_3CO_2Ag(s) + KBr(aq) \rightarrow AgBr(s) + CH_3CO_2K(aq)$

Silver bromide precipitate

d No reaction

e A reaction occurs

Ionic: $Ag^+(aq) + I^-(aq) \rightarrow AgI(s)$

Balanced: $AgNO_2(aq) + NaI(aq) \rightarrow NaNO_2(aq) + AgI(s)$

Silver iodide precipitate

f No reaction

g A reaction occurs

Ionic: $Pb^{2+}(aq) + SO_4^{2-}(aq) \rightarrow PbSO_4(s)$

Balanced: $Na_2SO_4(aq) + Pb(NO_3)_2(s) \rightarrow 2NaNO_3(aq) + PbSO_4(s)$

No precipitate formed

h A reaction occurs

Ionic: $Cu^{2+}(aq) + 2OH^-(aq) \rightarrow Cu(OH)_2(s)$

Balanced: $2N^-OH(aq) + Cu(NO_3)_2(aq) \rightarrow 2NaNO_3(aq) + Cu(OH)_2(s)$

Copper hydroxide precipitate

i No reaction

j Balanced: $Na_2S(aq) + Cd(NO_3)_2(aq) \rightarrow CdS(s) + 2NaNo_3(aq)$

Ionic: $Cd^{2+}(aq) + S^{2-}(aq) \rightarrow CdS(s)$

Balanced: $Na_2S(aq) + Cd(NO_3)_2(aq) \rightarrow CdS(s) + 2NaNo_3(aq)$

Cadmium sulfide precipitate

EVALUATION

1 C

2 C

3 Ionic: $Ca^{2+}(aq) + CO_3^{2-}(aq) \rightarrow CaCo_3(s)$
Balanced: $Li_2CO_3(aq) + Ca(NO_3)_2(aq) \rightarrow CaCO_3(s) + 2LiNO_3(aq)$
Calcium carbonate is formed

9780170412391

CHAPTER 18 SOLUBILITY

18.1 IMPORTANT TERMS

Across

1 kinetic
3 electrolyte
4 hydrated
6 dispersion
7 homogeneous
8 immiscible
9 intermolecular
10 exothermic
11 endothermic

Down

2 curve
4 hydrophobic
5 dissociation

18.2 FORMING SOLUTIONS

1 a D
b A
c D
d C
e E
f E

2 a Soluble – ionic bonding
b Insoluble – non-polar covalent bonding
c Soluble – polar covalent bond which ionises
d Insoluble – non-polar covalent bonding
e Soluble – polar covalent bond
f Soluble – ionic bonding

3 a $NH_4NO_3(s) \rightarrow NH_4^+(aq) + NO_3^-(aq)$
b $LiOH(s) \rightarrow Li^+(aq) + OH^-(aq)$
c $H_2SO_4(l) \rightarrow 2H^+(aq) + SO_4^{2-}(aq)$

4 Intermolecular bonding between water molecules is hydrogen bonding; between oil molecules is dispersion forces; between oil and water is dispersion forces. Therefore water/oil interactions are not sufficiently strong to disrupt H-bonds between water molecules.

18.3 SOLUBILITY CURVES

1

SOLUTION (IN 100 mL OF WATER)	SATURATED OR UNSATURATED?	IF UNSATURATED: HOW MUCH MORE SOLUTE CAN DISSOLVE IN THE SOLUTION (AT THE SAME TEMPERATURE)?
80 g of $NaNO_2$ at 20°C	Unsaturated	Approximately 8 g
40 g of NH_4Cl at 60°C	Unsaturated	Approximately 18 g
60 g of KNO_3 at 40°C	Unsaturated	Approximately 5 g
40 g of NaCl at 80°C	Saturated	N/A

2

SUBSTANCE	T (°C)	SOLUBILITY IN 100 g H_2O
$Pb(NO)_2$	20	55
CaCl	32	93
NH_3	10, 30, 80	70, 45, 13
NH_4Cl	50	52

3 a Less soluble
b Solid compounds become more soluble with temperature
c The dissolution of gases in water is an exothermic process, which is reversed increasingly as the temperature is raised.

EVALUATION

1 D
2 C
3 A
4 a

SOLUBILITY OF $K_2Cr_2O_7$	T (°C)	SOLUBILITY
Solubility (g per 100 g of H_2O)	40	20
Solubility (mol L^{-1})	40	
Solubility (g per 100 g of H_2O)	80	60
Solubility (mol L^{-1})	80	

b i Saturated (super saturated)
ii Unsaturated - approximately 5 g more will dissolve
iii Saturated

CHAPTER 19 pH

19.1 IMPORTANT TERMS

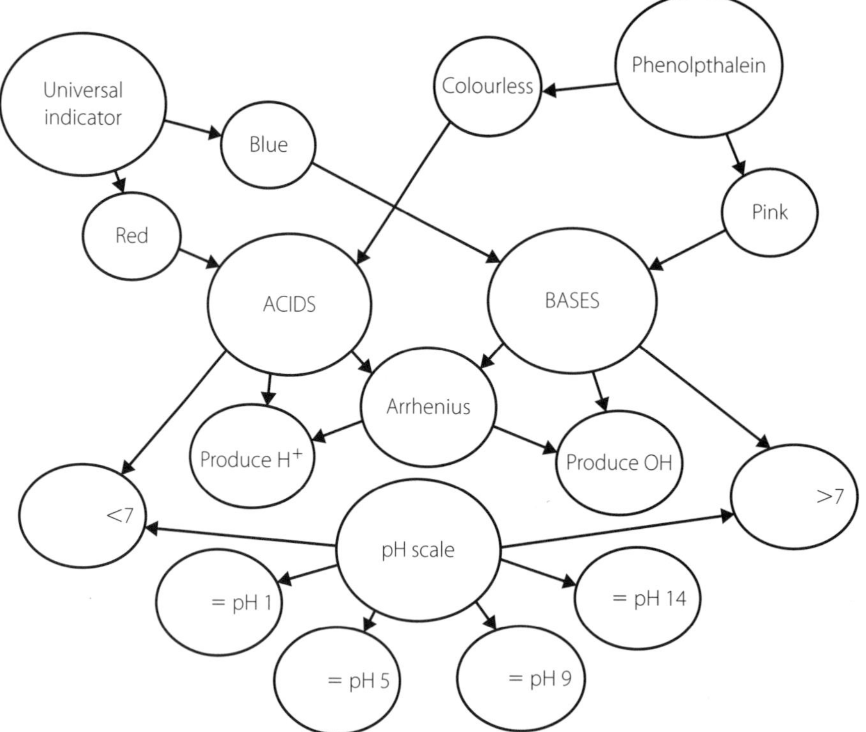

19.2 pH

1 a A substance that produces H^+ ions in solution.

b A substance that produces OH^- ions in solution.

c An amphoteric substance is one that can act as an acid or a base.

d A strong acid is one that dissociates completely to produce H^+ ions.

e A strong base is one that dissociates completely to produce OH^- ions.

f A weak acid is one that dissociates partially to produce H^+ ions.

g A weak base is one that dissociates partially to produce OH^- ions.

2 a A strong **base** is one that dissolves completely to produce OH^- ions in solution. Strong bases included **hydroxides and/or oxides** of groups I and II.

b A weak acid such as **acetic acid** only **partially** dissociates to produce **H^+ ions** in solution.

c An amphoteric substance is one that **acts as an acid or a base**.

19.3 THE pH SCALE

1 a A **strong base** would have a pH value of 13 whereas a weak **base** would have a pH of 8.

b A solution resulting from the reaction of the same amounts of acid and base would have a pH of **7**.

c An acid with a pH of 6 would release **10 000** times **less** H^+ ions in solution than a solution with a pH of 2.

19.4 THE ARRHENIUS MODEL

1 b methylamine: W

c ethanoic acid: W

d potassium hydroxide: S

e carbonic acid: W

f ammonia: W

g $C_8H_5KO_4$: W

EVALUATION

1 C

2 D

3 D

4 Amphoteric

5 An indicator

6 a potassium sulphate + water

b ammonium ethanoate (acetate) + water

c sodium potassium phthalate + water

7 a Substance R is a medium base and would therefore turn the indicator blue.

b Substance Q is a strong base and would therefore turn the indicator purple.

c Due to its reaction with zinc, O is an acid; its conductivity is low, therefore it will turn the indicator orange.

d Substance T is a weak base and would therefore turn the indicator turquoise.

9780170412391

CHAPTER 20 REACTIONS OF ACIDS

20.1 REACTIONS OF ACIDS AND BASES

1 a Hydrochloric acid + magnesium oxide → **magnesium chloride + water**

b **sulfuric acid** + magnesium carbonate → magnesium sulfate + **carbon dioxide** + **water**

c **acetic acid** + zinc → zinc acetate +**hydrogen**

d **phosphoric acid** + potassium hydrogen carbonate → potassium phosphate **carbon dioxide** + **water**

e Nitric acid + **ammonia** → ammonium nitrate.

2 a $2HNO_3(aq) + MgCO_3(s) \rightarrow Mg(NO_3)_2(aq) + CO_2(g) + H_2O(l)$

b $H_3PO_4(aq) + 3NH_3(aq) \rightarrow (NH_4)_3PO_4(aq)$

c $H_2CO_3(aq) + Zn(s) \rightarrow ZnCO_3(aq) + CO_2(g) + H_2O(l)$

d $H_2SO_4(aq) + 2KHCO_3(aq) \rightarrow K_2SO_4(aq) + CO_2(g) + H_2O(l)$

e $2CH_3COOH(aq) + MgO(s) \rightarrow (CH_3COO)_2Mg(aq) + H_2O(l)$

f $NH_4NO_3(aq) + (s) \rightarrow NaOH(aq) \rightarrow NaNO_3(aq) + NH_3(aq) + H_2O(l)$

20.2 INQUIRY SKILLS: CONSTRUCTING AND USING REPRESENTATIONS

1 a $2HNO_3(aq) + Na_2CO_3(aq) \rightarrow 2NaNO_3(aq) + CO_2(g) + H_2O(l)$

$2H^+(aq) + 2NO_3^-(aq) + 2Na^+(aq) + CO_3^{2-}(aq) \rightarrow 2Na^+(aq) + 2NO_3^-(aq) + CO_2(g) + H_2O(l)$

$2H^+(aq) + CO_3^{2-}(aq) \rightarrow CO_2(g) + H_2O(l)$

b $H_2CO_3(aq) + 2NH_3(aq) \rightarrow (NH_4)_2CO_3(aq)$

$2H^+(aq) + CO_3^{2-}(aq) + 2NH_3(aq) \rightarrow 2NH_4^+(aq) + CO_3^{2-}(aq)$

$2H^+(aq) + 2NH_3(aq) \rightarrow 2NH_4^+(aq)$

c $CH_3COOH(aq) + KHCO_3(aq) \rightarrow CH_3COOK(aq) + CO_2(g) + H_2O(l)$

$CH_3COO^-(aq) + H^+(aq) + K^+(aq) + HCO_3^-(aq) \rightarrow CH_3COO^-(aq) + K^+(aq) + CO_2(g) + H_2O(l)$

$H^+(aq) + HCO_3^-(aq) \rightarrow CO_2(g) + H_2O(l)$

d $2H_3PO_4(aq) + 3Mg(s) \rightarrow Mg_3(PO_4)_2(aq) + 3H_2(g)$

$6H^+(aq) + 2PO_4^{3-}(aq) + 3Mg \rightarrow 3Mg^{2+}(aq) + 2PO_4^{3-}(aq) + 3H_2(g)$

$6H^+(aq) + 3Mg \rightarrow 3Mg^{2+}(aq) + 3H_2(g)$

$2H^+(aq) + 3Mg \rightarrow 3Mg^{2+}(aq) + 3H_2(g)$

e $NaOH(aq) + NH_4Cl(aq) \rightarrow NaCl(aq) + NH_3(aq) + H_2O(l)$

$Na^+(aq) + OH^-(aq) + NH_4^+(aq) + Cl^-(aq) \rightarrow Na^+(aq) + Cl^-(aq) + NH_3(aq) + H_2O(l)$

$OH^-(aq) + NH_4^+(aq) \rightarrow NH_3(aq) + H_2O(l)$

EVALUATION

1 D

2 B

3 potassium sulphate

4 phosphoric acid

5 a

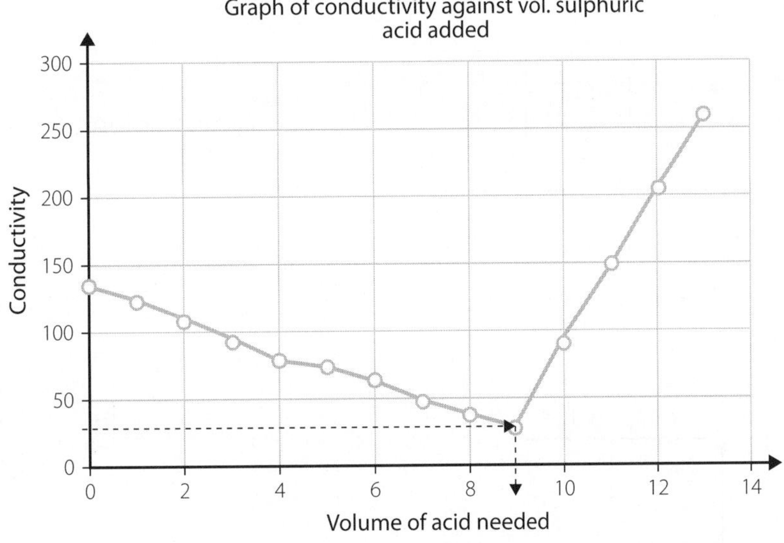

b endpoint occurs at 8.7–8.9 mL.

c As H_2SO_4 is added, it neutralises the $Sr(OH)_2$. Therefore, there are fewer ions in solution because some Sr^{2+} ions and some SO_4^{2-} are combining to form the $SrSO_4(s)$ precipitate. Eventually all of the strontium hydroxide has been used up. Adding more H_2SO_4 increases the conductivity.

CHAPTER 21 RATES OF REACTIONS

21.1 IMPORTANT TERMS

Across

3 Products

4 Activated

7 Collision

12 Maxwell-Boltzmann
13 Enthalpy
15 Pressure
16 Endothermic
18 Transition
20 Surface area
21 Exothermic

Down

1 Gradient
2 Activation
5 Concentration
6 Reaction rate
8 Nanoparticle
9 Kinetic
10 Catalyst
11 Temperature
14 Reactants
17 Energy
19 Rate

21.2 CONSTRUCTING ENERGY PROFILE DIAGRAMS

1

Energy

$S(s) + O_2(g)$

$\Delta H = -2296\ kJ\ mol^{-1}$

$SO_2(g)$

2

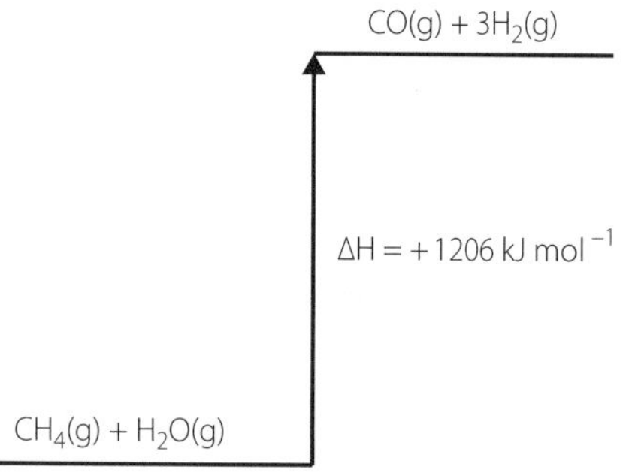

21.3 MODELLING SUCCESSFUL AND UNSUCCESSFUL COLLISIONS

1

COLLISION TYPE	DIAGRAM	RESULT
a Sufficient energy at the correction orientation		
b Correct orientation but insufficient energy		
c Sufficient energy but wrong orientation		
d Insufficient energy and wrong orientation		

21.4 EFFECT OF ALTERING CONDITIONS ON REACTION RATES

1

ACTION	EFFECT	EXPLANATION
a Heating reactants	Increase	Higher frequency of collisions. Higher proportion of successful collisions.
b Using a powdered form of a solid reactant	Increase	Greater surface area contact therefore higher frequency of collisions.
c Increasing the pressure on a gas reactant	Increase	Higher concentration of gas particles so higher frequency of collisions.
d Cooling a liquid reactant	Decrease	Lower rate of movement of gas particles, so lower frequency of collisions.
e Increasing the concentration of a liquid reactant	Increase	Higher frequency of collisions.
f Adding more water to dissolved reactants	Decrease	Lower concentration, so lower frequency of collisions.
g Adding a catalyst to dissolved reactants	Increase	Lower activation energy for reaction.

9780170412391

ACTION	EFFECT	EXPLANATION
h Increasing the volume of a gas reactant	Increase	Increase in number of successful collisions.

21.5 GRAPHING AND INTERPRETING RESULTS

1 To trap gas inside the cylinder – gas cannot pass through glass. Water enables amount of gas to be measured.

2 Turns limewater cloudy

3 Reactants: $CaCO_3$, HCl

Products: CO_2, H_2O, $CaCl_2$

4 calcium carbonate + hydrochloric acid → calcium chloride + carbon dioxide + water

$CaCo_3(s) + 2HCl(aq) \rightarrow CaCl_2(aq) + CO_2(g) + H_2O(l)$

5

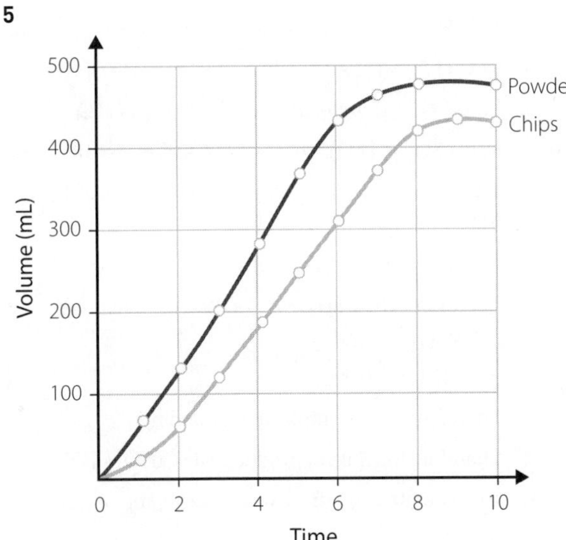

6 Rate is highest at start and decreases as the reaction completes

7 Powder

8 Surface area of powder is greater than that of chips, therefore rate of reaction is higher

9 Reaction finished earlier

10 In slower reaction, more CO_2 dissolved in the water, therefore was never collected as gas

EVALUATION

1 a, 2a

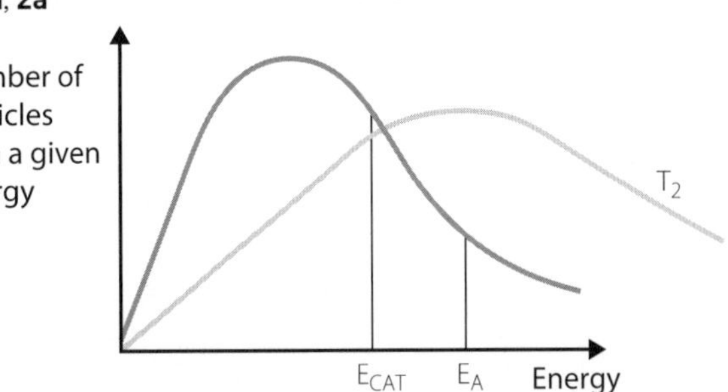

1 b Enables the reaction to occur via a different pathway with a lower activation energy. Therefore a higher number of particles have sufficient energy to react and the reaction is faster.

2 b Higher temperature causes particles to move more quickly, therefore greater frequency of collisions and greater proportion of fruitful, high energy collisions leading to a reaction.

3 Change in concentration leads to a greater frequency of collisions, but concentration does not affect the energy of the particles, therefore cannot be represented on the diagram.

PRACTICE EXAMINATION ANSWERS

MULTIPLE CHOICE

1 C

2 B

3 B

4 D

5 C

6 C

7 B

8 D

9 B

10 B

SHORT ANSWER AND COMBINATION-RESPONSE QUESTIONS

1 Reaction **a** is endothermic

Reaction **b** is exothermic

2 The isotopes are **A** and **J**

ATOM	ATOMIC NUMBER	MASS NUMBER	NUMBER OF NEUTRONS
A	23	56	**33**
G	22	55	**33**
J	23	55	**32**

3 Two additional properties: sour tasting and reacts with metals.

4

PROPERTY	INCREASE, DECREASE OR UNCHANGED?
Temperature	**Increase**
Volume	**Unchanged**
Pressure	**Increase**
Kinetic energy of the gas particles	**Increase**

5 $1s^2\ 2s^2\ 2p^6\ 3s^2\ 3p^6\ 4s^1\ 3d^5$

6 Average mass of a molecule relative to ^{12}C

7 Particle size – smaller particles are carried more quickly

Adsorption to stationary phase – particles that adsorb more are carried more slowly

8 **a** Intermolecular forces – the greater the force, the higher the boiling point.

b Intermolecular force is dispersion forces. The greater the mass of the molecules, the greater the dispersion forces, so the higher the boiling point.

c CH_4 has dispersion forces but NH_3, HF and H_2O all have hydrogen bonding forces between molecules as well as dispersion forces, so boiling points are significantly higher.

d NH_3 has an additional lone pair of electrons, therefore the shape is trigonal pyramid, whereas BCl_3 is trigonal planar.

9 When heated they absorb energy which causes electrons to move to a higher energy state. The excited electrons then fall back to their ground state, releasing energy of a specific wavelength and therefore colour.

HIGHER-ORDER QUESTIONS

1 **a** **i** C 54.6% $54.6 \div 12 = 4.55$ $4.55 \div 2.27 = 2$

H 13.6% $13.6 \div 1 = 13.6$ $13.6 \div 2.27 = 6$

N 31.8% $31.8 \div 14 = 2.27$ $2.27 \div 2.27 = 1$

C_2H_6N

ii $M(C_2H_6N) = ((2 \times 12) + (6 \times 1) + 14) = 44g$

$88 \div 44 = 2$, therefore molecular formula is $C_4H_{12}N_2$

b **i** $M_r(CuSO_4 \bullet 5H_2O) = 63.5 + (32.1 \times 4) + (5 \times 18) = 281.9$

Therefore n (original compound) $= 5.838 \div 281.9 = 0.0207$ moles

ii m(water removed) $= 5.838 - 4.153 = 1.685$ g

Therefore n (water) $= 1.685 \div 18 = 0.0936$ moles

iii $\dfrac{n(H_2O)}{n(CuSO_4)} = \dfrac{0.0936}{0.0207} = 4$

Therefore 1 water molecule remaining

Therefore formula of product was $CuSO_4 \bullet H_2O$

iv To ensure that no more water was being removed.

2 **a** **i** Increases

ii Nuclear charge increases, therefore attraction between nucleus and electrons increases, therefore electron is harder to remove

iii Decreases

iv Increased number of electron shells, therefore increased shielding of electrons from nuclear attraction, and electrons are easier to remove

b **i** Sodium is more metallic than sulfur (conducts electricity as a solid)

ii Non-metallic

c RAM =

$$\frac{(46 \times 7.93) + (47 \times 7.28) + (48 \times 73.94) + (49 \times 5.51) + (50 \times 5.34)}{100}$$

$= 44.82$

Scandium

3 **a** $2H_2O_2(aq) \rightarrow O_2(g) + 2H_2O(l)$

b **i** $n = \dfrac{Pv}{Rr} = \dfrac{99.9 \times 0.476}{8.31 \times (273 + 22.6)} = 0.0194 \text{moles}$

ii $n(H_2O_2) = 2 \times n(O_2)$

$= 0.0388$ moles

Therefore $C(H_2O_2) = 0.0388 \div 0.05 = 0.774$ M

9780170412391

c Reaction proceeds via a different pathway with lower activation energy, so greater proportion of particles are able to react.

4 a $P_1V_1 = P_2V_2$ $\frac{V_1}{V_2} = \frac{6}{7}$

Therefore $P_2 = P_1 \times \frac{V_1}{V_2} = 254 \times \frac{6}{7}$

$= 217.7$ kPa

b $\frac{P_1V_1}{T_1} = \frac{P_2V_2}{T_2}$ therefore $V_2 = \frac{P_1V_1 \times T_2}{T_1 \times P_2}$

$= \frac{100.6 \times 400 \times 233}{295 \times 4.18}$

$= 7604$

5 a Hydrogen bonding intermolecular forces between water molecules. Dispersion forces between methane molecules are much weaker. Therefore less energy is required to separate liquid methane molecules to form a gas.

b **i** $C = 159.5 \div 5 = 31.9$ g per 100 g

ii Unsaturated (less than 36.2)

6 a Two from:

B Reactants lose energy as heat to form products

C Reactants in an endothermic reaction absorb heat in order to form products

E Temperature increases the rate of all reactions

b Two from:

A Incorrect – some exothermic reactions are not spontaneous and some endothermic reactions are spontaneous. Activation energy is not dependant on enthalpy change.

D Catalysts change reaction pathway and affect activation energy, not enthalpy change.

F Rates of all reactions decrease as they proceed.

c **i** $Zn(s) + 2HCl(aq) \rightarrow ZnCl_2(aq) + H_2(g)$

ii Finely grained zinc powder – greater surface area contact, and higher frequency of collisions possible.